spatial data analysis

For Gemma

spatial data analysis

AN INTRODUCTION FOR GIS USERS

Christopher D. Lloyd

School of Geography, Archaeology, and Palaeoecology
Queen's University, Belfast

OXFORD
UNIVERSITY PRESS

OXFORD
UNIVERSITY PRESS

Great Clarendon Street, Oxford OX2 6DP

Oxford University Press is a department of the University of Oxford.
It furthers the University's objective of excellence in research, scholarship,
and education by publishing worldwide in

Oxford New York

Auckland Cape Town Dar es Salaam Hong Kong Karachi
Kuala Lumpur Madrid Melbourne Mexico City Nairobi
New Delhi Shanghai Taipei Toronto

With offices in

Argentina Austria Brazil Chile Czech Republic France Greece
Guatemala Hungary Italy Japan Poland Portugal Singapore
South Korea Switzerland Thailand Turkey Ukraine Vietnam

Oxford is a registered trade mark of Oxford University Press
in the UK and in certain other countries

Published in the United States
by Oxford University Press Inc., New York

British Library Cataloguing in Publication Data

Data available

Library of Congress Cataloging in Publication Data

Lloyd, Christopher D.
   Spatial data analysis : an introduction for GIS users / Christopher D. Lloyd.
      p. cm.
   ISBN 978-0-19-955432-4
1. Geography—Statistical methods. 2. Geographic information systems. I. Title.
   G70.3.l56 2010
   910.01'5195—dc22

                                                                    2009039593

Typeset by Macmillan Publishing Solutions
Printed in Italy
on acid-free paper by
L.E.G.O. S.p.A.

ISBN 978-0-19-955432-4

1 3 5 7 9 10 8 6 4 2

■ Preface

There is a plethora of books available that introduce, at a variety of levels, approaches for the analysis of spatial data. Why write yet another one? The most obvious reason is that, while many of the existing books provide excellent accounts of a wide range of methods, none of the existing texts started and concluded at a level that I felt was appropriate for the students (mostly at an undergraduate level) whom I have had the responsibility to teach. Many existing introductory texts on geographical information systems (GIS) include material at an appropriate level about data models, databases, visualization, and elements of spatial analysis, amongst other themes. However, in my experience, a combination of texts was required to address fully the key issues in spatial data analysis in sufficient depth. While some texts covered many approaches to spatial data analysis, the prior expectations, in terms of existing knowledge, made such texts inaccessible to many undergraduate students in geography and others with a burgeoning interest in spatial data analysis. This book is intended as a bridge between texts that introduce some key principles of GIS and those that go into a great deal of depth about a wide body of methods for spatial analysis.

Another concern is to introduce in a reasonably small amount of text a set of key ideas so that the book could be used to support part of a course concerned with, for example, GIS and GIScience in general. Although there is a focus on how the methods work, the aim is also to develop an understanding of when and how it is appropriate to apply particular methods. One key reason for writing this book follows from my belief that academic courses in GIScience should focus on education and not simply on training—that is, while it is important for students to learn how to use software, without some understanding of what is being done and of the implications of particular decisions, the process is reduced to training.

This book is also a logical progression from previous work: another book (Lloyd, 2006) published before this book was started had a rather different focus and intended audience with some introductory material provided but a fair amount of background knowledge assumed. On finishing that book, writing a more general text that started from first principles was tempting and the present book is the result. I hope that this book builds knowledge and fosters interest. If it provokes readers to find out more then it will have succeeded in its aims.

■ Acknowledgements

The case studies in this book are based on data from a variety of sources. The British Atmospheric Data Centre (BADC) is thanked for provision of United Kingdom Meteorological Office (UKMO) Land Surface Observations Data. The UKMO is acknowledged as originator of these data. The Northern Ireland Statistics and Research Agency (NISRA) is acknowledged for providing access to data from the 2001 Northern Ireland Census of Population. Census output is Crown copyright and is reproduced with the permission of the Controller of HMSO and the Queen's Printer for Scotland (under the terms of the Click-Use Licence). Gregoire Dubois is thanked for making available the SIC97 dataset through the AI-Geostats website.

I am grateful to Gemma Catney and Jenny McKinley for their comments on parts of the text. The anonymous reviewers of the original proposal and of earlier versions of the text are thanked for their helpful comments. I have endeavoured to take all comments into account but any errors or inaccuracies which remain are the responsibility of the author.

Contents

Introduction

1.1 Spatial data analysis

Geographical or spatial data play a vital role in many parts of daily life. Either directly, as in the use of a map for navigating around a city, or indirectly, where we use resources like water or gas, we are dependent on information about where things are located and about the attributes of those things. Making use of spatial data requires a whole set of approaches to extract information from those data and make them useful. Geographical information systems (GIS) play a key role in this context. GIS provide a means of generating, modifying, managing, analysing, and visualizing spatial data. The key contribution of GIS, above and beyond functions provided by other forms of software such as cartographic mapping or computer-aided design packages, is in the analysis of spatial data. Broadly, analysis is concerned with breaking apart a problem with the aim of finding a solution to this problem. In terms of this book, the aim is to introduce some ideas and methods that may be useful in the analysis of spatial data. For example, approaches are presented for:

- summarizing a set of values (e.g. what is the mean average of all values?)
- identifying overlaps between different features (e.g. what areas with pollution above some critical threshold are located in areas that have a population of greater than a particular size?)
- finding the shortest route between one place and another through a network (e.g. a road network)
- identifying clustering in point events such as cases of some disease (e.g. where are disease incidence rates highest?)
- exploring spatial patterning in variables (a variable being a quantity that may vary across samples; precipitation amount or elevation, for example, can be considered variables) (e.g. does the concentration of some pollutant vary spatially and where are values largest?)

- exploring how two or more variables are related at different places (e.g. does the relationship between altitude and snowfall vary from place to place?)
- estimating the values of some property (e.g. precipitation amount) at locations where there are no samples available (necessary as a prior stage to many other procedures)
- assessing the construction costs of alternative routes for a new road.

There are many other kinds of approaches covered in this book, but many are based on common concepts such as measurement of distances or differences in properties in different areas. These fundamental concepts are described along with some particularly widely used approaches and the selected approaches are illustrated with example applications. The emphasis throughout is on *education* rather than simply *training*, based on the conviction that users of spatial data analysis tools should know something about how the approaches work rather than simply how to apply them. Appendix G details some spatial analysis tasks and the sections of the book that contain relevant material. The remainder of this chapter sets out the purpose of the book and its contents.

1.2 Purpose of the book

The aim of this book is to introduce a set of key ideas or frameworks that will give the reader knowledge of the kinds of problems that can be tackled using widely available tools for the analysis of spatial data. Another key concern is that readers *understand* how the methods work, therefore a large majority of methods introduced are demonstrated through small case studies. The book includes detailed coverage of a relatively limited number of basic key methods for the analysis of spatial data. These are intended to illustrate the workings of particular methods as well as to demonstrate key concepts that will support understanding of other approaches.

This is not an introduction to GIS. Readers who wish to know about, for example, data models, databases, or visualization will find brief introductions in this book, although they should consult one of a range of textbooks such as those by Heywood *et al.* (2006) or Longley *et al.* (2005a) for more in-depth accounts. A full description of some key GIS algorithms (put simply, sets of instructions that are worked through to achieve some particular objective) is provided by Wise (2002). In this book, little prior knowledge is assumed of readers. However, it is expected that readers will have some basic knowledge of GIS principles. The book is also not an introduction to statistics, although some key ideas are discussed. The book by Rogerson (2006) provides a good introduction to statistics for geographers. Only a very limited prior knowledge of statistics is required to make full use of the present book and it is an aim in this book that no terms likely to cause confusion will be dropped in without explanation. While it is assumed that most readers will work through the book systematically, it is designed so that it is possible to dip in to a particular topic and the discussions about methods are, in the majority of cases, fairly self-contained.

Many other books provide introductions to spatial data analysis (or geospatial analysis, as it sometimes called). The book by O'Sullivan and Unwin (2002) provides an excellent account of some key ideas as well as more advanced material. Lloyd (2006), De Smith *et al.* (2007), and Chang (2008) provide other detailed accounts. Lee and Wong (2000) also provide clear introductions to core principles and in-depth accounts of more complicated ideas. The book by Cressie (1993), which focuses on spatial statistics, is encyclopaedic in its coverage, extremely well regarded, and has become something of a standard work. Other works, such as the book by Griffith (1988), seek to disseminate research findings for those already familiar with the principles of spatial data analysis. The present book is intended to have a rather different focus to any of these books. The specific intention is to communicate some ideas and concepts that are central to the analysis of spatial data without discussing alternative approaches in great depth. It is hoped that the book will provide material to develop the reader's conceptual understanding such that books dealing with a broader range of methods can then be encountered with greater confidence. By keeping things to the point, but addressing key issues, I hope that this book will build knowledge and interest and encourage enthusiasm for learning more about spatial data and their analysis.

As noted previously, it is assumed that readers will have some background knowledge of GIS. Reference is made to key principles such as data models and databases with the expectation that these will be familiar. However, an attempt is made to introduce all potentially new topics. The book only briefly discusses key topics such as data input, visualization, errors and error propagation (transfer of errors from one processing stage to another). Instead, the focus is directly on the analysis of spatial data and, where appropriate, other sources of information are suggested. The further reading section at the end of this chapter provides some starting points for major issues that are not discussed in depth here and each chapter has a further reading section at its end.

1.3 Key concepts

A book of this length can only provide an in-depth account of a limited range of methods and ideas; it necessarily skirts over many major issues. Nonetheless, the aim has been to provide sufficient background to some key concepts and some specific approaches that readers will be able to explore other methods in an informed way. The next three chapters seek to outline a set of basic principles, an understanding of which is necessary to make use of the rest of the text. For readers new to the topic of spatial data analysis these chapters may be more challenging than most of the rest, but it is hoped they will allow development of a knowledge base appropriate for making use of the rest of the book.

1.4 Structure of the book

Chapters 2, 3, and 4 introduce a range of key concepts that provide the foundations for the rest of the book. Chapter 2 describes data models (ways of representing real-world objects or features), data management, spatial scale, data collection, data errors, visualization, and querying. Chapter 3 introduces some key statistical concepts and methods. In Chapter 4, some methods for the analysis of spatial data are outlined. More specifically, Chapter 4 deals with measuring distances and areas, moving windows (a key concept in spatial data analysis, which allows assessment of differences between geographical areas), geographical weights, and a variety of additional core issues in the handling and analysis of spatial data. Chapter 5 is concerned with overlaps between features (e.g. do areas with particular characteristics overlap?) while Chapter 6 is concerned with links between component parts of networks (e.g. a road network) to address questions like what is the shortest path between locations A and B? Chapter 7 introduces some methods for the analysis of point patterns (geographically located sets of point events such as cases of a disease). Chapter 8 is concerned with the analysis of spatial patterning in single variables and in relations between multiple variables. In simple terms, it presents methods for exploring how values vary geographically and how the relationships between values vary (the example given in Section 8.7 considers the relationship between altitude and snowfall). Chapter 9 outlines some methods for generating surfaces from point data and for transferring values between different sets of zones. In Chapter 10, the focus is on the analysis of grids and surfaces (both literally, as in topography, and in terms of other properties, such as precipitation, that can be treated as surfaces). Finally, Chapter 11 pulls together some key themes addressed in the book and suggests some ways forward for those who would like to know more about the topics addressed in the book.

Appendices A to F include short outlines of particular topics to support discussion in the main body of the text and these are referred to where relevant. Appendix G includes a table that details some common kinds of problems that are often encountered in spatial data analysis. The relevant sections of the book that offer solutions to these problems are detailed. The table is intended as a means of quickly identifying sections of the text that are relevant for specific applications.

All of the substantive chapters include worked examples using either synthetic or real-world data. In addition, Chapters 5 to 10 include case studies at their conclusions with the data on which these studies are based being provided on the book website along with guidance on how some key methods are implemented in popular GIS packages. All of the synthetic data are also provided on the book website. It is hoped that the text, example applications, and data will, in conjunction, allow readers to develop a firm understanding of the key ideas and techniques described in the book.

Further reading

Each chapter details some texts that provide more detail on the topics discussed. In this case, some introductions to GIS are suggested. General introductions with descriptions of data models, data input procedures, and spatial data management include the books by Burrough and McDonnell (1998), Longley *et al.* (2005a), and Heywood *et al.* (2006). The books edited by Longley *et al.* (2005b) and Wilson and Fotheringham (2008) provide detailed accounts of key issues and concepts. There are also several general introductions to GIS written for specific disciplines or research areas. These include books for geoscientists (Bonham-Carter, 1994), social scientists (Martin, 1996; Steinberg and Steinberg, 2006) and archaeologists (Conolly and Lake, 2006).

➡ The following chapter introduces some key concepts in GIS and is the first of three chapters that outline some fundamental principles on which the rest of the book builds.

2
Key concepts 1
GIS

2.1 Introduction

This chapter is the first of three intended to introduce a set of key concepts that are central to the approaches described in the rest of the book. This chapter focuses on some key principles of GIS and is followed by chapters on some fundamentals of statistics and spatial data analysis. This chapter will briefly discuss ways in which real-world entities can be represented in a GIS. The focus will then move on to methods for spatial data collection, errors, visualization of spatial data, and simple approaches for extracting information from data sets. More specifically, the chapter covers:

- data and data models, including topology—representing reality in a GIS using data models and databases
- spatial referencing systems and projections—how spatial data are spatially referenced and how parts of the Earth's surface can be represented on a map
- spatial scale—how far spatial properties vary over a given area
- data collection—how spatial data are generated
- sources of errors in spatial data
- visualization of spatial data—a key first step in any analysis
- querying spatial data—extracting information from spatial databases.

The following section deals with data models used commonly in GIS.

2.2 Data and data models

A model is simply a means of representing 'reality' and spatial data models provide abstractions of spatially referenced features in the real world. The focus of this book is on analysis of spatial data rather than the ways in which spatial data are structured. However, a very brief introduction to data models was considered useful as knowledge of analysis of spatial data requires at least a basic understanding of data structures. Most books on GIS stress the division of data models into the well-known raster and vector representations. The key properties of the two data models, which will be useful in making sense of the rest of this book, are outlined here. The data model available determines the choice of method for spatial analysis as do the characteristics of the particular data set.

Representations of real-world features are often divided into (1) entities and (2) fields (Burrough and McDonnell, 1998). Entities are conceptually distinct objects like point locations, roads, or administrative boundaries. Fields convey the idea of values of some property at all locations. For example, elevation can be measured or estimated at all places and elevation does not usually have distinct edges, in contrast with, for example, buildings. Objects that are well described as distinct entities are sensibly represented using the vector data model. Properties that tend to vary quite smoothly from place to place (i.e. they are spatially continuous and their values do not tend to change abruptly from place to place) are frequently represented using the raster data model. There are notable exceptions and these include isolines and contours, which are vector-based representations of continuous phenomena such as temperature or elevation (there are, of course, exceptions—a cliff edge represents an abrupt change in elevation and so temperature is perhaps a more conceptually straightforward example). The raster and vector data models are briefly defined in turn. A more in-depth account of the way in which information is stored using the two data models is given by Wise (2002).

2.2.1 Raster data

While it is assumed that readers are familiar with raster grids, some key issues are addressed here. Raster grids are conceptually simple structures, comprising square cells with numeric values or classes attached to each cell. A simple example of a raster grid is given in Figure 2.1; in this case the value represents elevations in metres. Where the cells contain categorical or integer (i.e. whole number) values the number of instances of each class may be stored in a table. In cases where values with decimal places are used, all information is conventionally stored in the raster itself. There are huge amounts of data available in raster grid format—remotely sensed imagery (see Section 2.8.3) comprises a major component of such data sources. The spatial resolution of a raster refers to the area in the real world covered by a cell. For example, a grid with a spatial resolution of 5 m covers an area of 5 by 5 (=25) square metres. Remotely sensed images with very fine spatial resolutions (e.g. 1 m) have been generated for many parts of the world, although ease of access (cost, etc.) varies geographically.

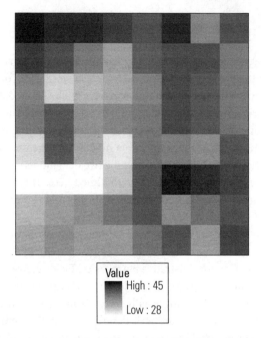

Figure 2.1 Raster image used to illustrate various methods. Units are elevations in metres.

2.2.2 Vector data

As with the raster model, it is assumed that readers are familiar with the vector data model, but some background is given about vector data storage formats and their relevance for spatial data analysis. Whereas features in raster grids are identified simply by the row and column position of cells, vector data comprise explicit spatial coordinates of the features that make up objects. Vector data comprise points (with x- and y-coordinates), lines (line segments (or arcs) connected by points), and area polygons (lines with the same start and end point). An example of some line features is given in Figure 2.2. This example shows that line features comprise two forms of point locations—vertices, which represent change in direction of arcs, and nodes, which represent the start or end of arcs, including locations where different arcs connect. Of course, vectors representing real-world features are usually much more complex than those shown in Figure 2.2 (and may have many vertices). Note that, while x and y are used to represent two-dimensional position, z is often used to indicate the third dimension (elevation) and also, as in this book, values of any property (e.g. precipitation amount) that are associated with particular x- and y-coordinates.

Vector data can be stored as what are sometimes called 'spaghetti' data—that is, strings of unconnected line segments. In this case, relationships between objects (e.g. which line is connected to which other lines) are not encoded. However, explicit information on the relationships between objects reduces the computational demands of subsequent analyses and analysis of vector data is usually preceded by the construction of topology, as discussed in the next section. Conventionally, vector data are divided

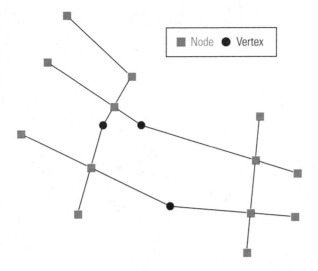

Figure 2.2 Vector line features used to illustrate network analysis methods in Chapter 6.

into their spatial component and their attribute component. Attributes linked to each spatial feature are often stored using a relational database system (see Section 2.3 for more on databases) and this is demonstrated in the following subsection.

2.2.3 Topology

This section is concerned with vector representation of objects and connections between features. Topology can be defined as 'the mathematical study of objects which are preserved through deformations, twistings and stretchings. (Tearing, however, is not allowed.)' (Weisstein, 2003, p. 2990). In other words, if a map showing a set of zones is stretched, zones separated by other zones cannot become neighbours. For this to be possible, the map would have to be cut (or torn as the quote above indicates) or folded over on itself (Wise, 2002). In a GIS (as well as a computer-aided design system, etc.), information on the connections between objects and, where appropriate, neighbouring objects, may be stored. Operations concerned with connections between objects (e.g. administrative areas or roads) are dependent on information about topological relationships.

Obviously, to represent a point only the point coordinates are required. In Figure 2.3, an example of line topology representation is given. In this case, arcs are indicated with the prefix 'A' and the nodes with the prefix 'N'. Changes in the direction of arcs between nodes are represented by vertices, as shown previously in Figure 2.2. Note from the table that direction is represented—there is a 'from node' and a 'to node'. N3 may, at first sight, appear to be a vertex but it connects two distinct arcs—A1 and A2.

Figure 2.4 gives an example of vector polygon topology representation. With this representation, information on arcs and the polygons to which they belong is contained in one table while information on the area and perimeter of the polygons is contained in another table. The table at the top has separate codes for each arc, the

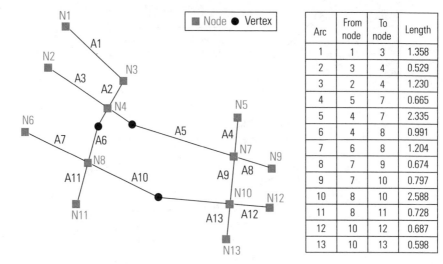

Arc	From node	To node	Length
1	1	3	1.358
2	3	4	0.529
3	2	4	1.230
4	5	7	0.665
5	4	7	2.335
6	4	8	0.991
7	6	8	1.204
8	7	9	0.674
9	7	10	0.797
10	8	10	2.588
11	8	11	0.728
12	10	12	0.687
13	10	13	0.598

Figure 2.3 Line topology representation. The prefix 'N' refers to node and the prefix 'A' to arc.

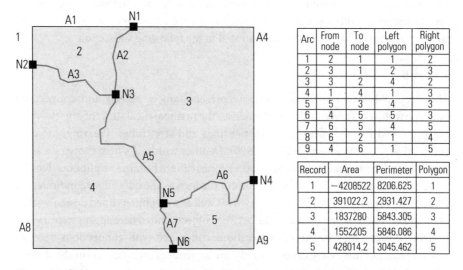

Arc	From node	To node	Left polygon	Right polygon
1	2	1	1	2
2	3	1	2	3
3	3	2	4	2
4	1	4	1	3
5	5	3	4	3
6	4	5	5	3
7	6	5	4	5
8	6	2	1	4
9	4	6	1	5

Record	Area	Perimeter	Polygon
1	−4208522	8206.625	1
2	391022.2	2931.427	2
3	1837280	5843.305	3
4	1552205	5846.086	4
5	428014.2	3045.462	5

Figure 2.4 Polygon topology representation. The polygon numbers are those without a prefix. The prefix 'N' refers to node and the prefix 'A' to arc.

start and end node (indicated with the prefix 'N' on the figure) of each arc (given by the prefix 'A' on the figure) and the polygon that lies to the left and right of each arc (given by numbers without a prefix). Note that the negative area for polygon 1 corresponds to the area outside the edge of the region and it is the negative of the sum of areas within the boundary (i.e. if you take the absolute value of this first area figure, this gives the total area within the outer edge). Additional tables with the same arc identifiers can be added easily with a relational database structure. With a relational database, each object has a unique identifier (a 'primary key'). If information in other tables is available, this information can be linked together if both sets of data use the same identifiers.

With this representation, information is recorded on connectivity (arcs are connected by nodes), containment (enclosed polygons can be identified, although there are none in the example), and contiguity (arcs shared between polygons, thus determining contiguity, are indicated, e.g. polygons 3 and 4 share arc 5). The name of the well-known software environment ArcInfo™ (now part of ArcGIS™) reflects the relationship between arcs and attribute information.

2.2.4 Rasters and vectors in GIS software

Most GIS environments allow for conversion from rasters to vectors or from vectors to rasters. Clearly, the spatial resolution of the raster will limit the positional accuracy of vector objects generated through raster to vector conversion. For many standard operations (e.g. measuring proximity to objects; see Section 4.5) either vectors or rasters can be used and so the choice of data model may, for many applications, be less important than it was in formally less flexible GIS software environments.

2.3 Databases

GIS are often defined as having at their core a spatial database. In essence, a database is simply a set of structured information. In order to store, manage, and access computer-based data a variety of database structures have been developed. The most frequently encountered is based on the relational database structure introduced by Codd (1970). The topological structures discussed in the previous section are based on tables linked by common identifiers, and this is the basis of a relational database. For example, in the case of the line topology representation in Figure 2.3, the arcs themselves and both tables have a unique arc number—there is only one arc with the label '1'. The information in the two tables can easily be joined together using this common identifier or key. In the case of the polygon topology representation in Figure 2.4, there are unique arc and polygon numbers. The polygons listed in the table at the top correspond to the polygons listed in the table at the bottom and the two tables can be linked using the polygon identifiers. One objective in setting up a relational database is to store only necessary information and to enable efficient and effective access to the data. The setting up of such a database may be non-trivial if there are multiple tables linked in different ways. While the relational database structure is encountered most frequently, other database structures are widely used. Object-oriented databases are quite commonplace; such systems organize data in a structured hierarchy. One example is the nesting of an individual within a house and a house within a census area and a census area within a larger administrative unit. Efficient organization of data is important for spatial data analysis (particularly in the case of large data sets) as speed of processing may, in many cases, be markedly increased if the database is well structured. Burrough and McDonnell (1998) and Worboys and Duckman (2004) provide more detailed descriptions of database systems. Conventionally, attributes of vector spatial data are often stored using a relational database system. Raster data are often stored as self-contained objects. However, as noted before, in cases where a raster grid is composed of integer values, an associated table may be used that records the number of instances of each value.

2.3.1 Database management

Data stored in a database are accessed using a database management system (DBMS). A DBMS offers facilities for updating the database and extracting data in a flexible way. The DBMS includes facilities to import data into the database, to manage user access, to update the database structure and content, and to conduct queries to extract specific information. The querying of spatial databases is the subject of Section 2.11. A good summary of DBMS is provided by Longley *et al.* (2005a).

2.3.2 The Geodatabase

The term 'Geodatabase' is sometimes used to mean a spatial database in general. More specifically, the Geodatabase is the core means of storing and managing spatial data in the ArcGIS™ environment. The Geodatabase offers various benefits over conventional GIS file formats such as shapefiles or coverages (both widely used and well-established vector file formats). In particular, the Geodatabase combines data into a single integrated database rather than storing each layer in discrete files. The Geodatabase has built-in rules which help to maintain the integrity of the database and reduce database maintenance.[1]

2.4 Referencing systems and projections

The location of spatial objects is usually recorded using some kind of spatial referencing system such as longitudes and latitudes, or eastings and northings using some kind of national grid system. Given that the surface of the Earth (as well as other bodies) is not flat, one of a variety of projection systems can be used to transform locations on a sphere to features on a flat surface. It is essential that data are projected appropriately, and a brief summary of some key issues is provided for context.

A projection can be defined as comprising the ellipsoid, the datum, and the projection. The ellipsoid is the smooth approximate shape of the Earth and an ellipsoid can be selected which best represents the surface of the Earth for the whole planet or for a given area. The surface which represents deviations from the ellipsoid is termed the 'geoid' (Clarke, 1999). The datum is the origin or centre and rotation of the ellipsoid.

---

1 http://www.esri.com/news/arcuser/0701/migrating.html

The datum is defined such that it has the best fit to the surface of the Earth over the area of interest.

Any projection entails making compromises, for example some projections distort areas while others distort distances. Conformal projections preserve shape while equal area projections preserve areas (as their name suggests) and not shapes—clearly a projection cannot preserve both properties. A common example of a conformal projection is the Mercator projection and an example of an equal area projection is the Albers equal area projection. Conformal projections like the Mercator projection preserve angles locally and thus the Mercator projection is well-suited to the purpose of navigation, for which it was developed. Projections also exist which preserve distances along one or several lines (Clarke, 1999). An awareness of projection systems is important in working with spatial data. If, for example, the concern is with areas of countries then an equal area projection must be used. The impact of choice of projection is a function of the spatial scale of the map. In short, the larger the area of concern, the greater the impact of selection of an appropriate projection will be (Clarke, 1999) and the impact of poor choice of projection when using a map with a representative fraction (i.e. ratio of distances on a map to distances in the same units in the real world) of 1:1,000,000 will be greater than when using a 1:10,000 map. A brief introduction to projections is provided by Longley *et al.* (2005a) while a more detailed account is given by Seeger (2005). Scale is discussed in Section 2.7.

2.5 Georeferencing

To make spatial data useable, they must be linked to some kind of spatial referencing system, as detailed above. The process of attaching spatial information to data is called georeferencing. An example of georeferencing is the use of coordinates obtained using a global positioning system (GPS) receiver to link positions on a remotely sensed image to positions that have been surveyed, and the linked survey points are called ground control points (GCPs). The process of transforming an image, whereby the transformed image fits well to the GCPs and the image is then in coordinate space, is termed 'georectification'. Georectified images can be overlaid or combined with other data that are georeferenced using the same system. The term 'geocoding' is generally used in relation to the determination of geographic coordinates from an address or related data (McDonnell and Kemp, 1995) and this is outlined next.

2.6 Geocoding

An example of geocoding is the conversion of addresses into geographic coordinates. A variety of databases and software environments exist to facilitate links between names, addresses, postcodes or zip codes, and geographic coordinates. Such data sources and tools may be part of an essential prior step to spatial analysis. In some contexts, information may be available on road networks and segments of properties that are located on the roads, but not on the specific location of properties. In such cases, given this knowledge, the locations of individual properties can be predicted through address interpolation (or address matching). The concept is illustrated in Figure 2.5, where the coordinates of the road junctions are known and the positions of properties in the segment are interpolated given that odd-numbered houses are located on the eastern side of the road while even-numbered houses are located to the west. The interpolated position of number 16 is given. Of course, such an approach may give misleading results where the size of properties, or the spacing between them, varies. An introduction to geocoding and related issues is provided by Longley *et al.* (2005a).

2.7 Spatial scale

Spatial data analysis is dependent on the sample size, density, and, where relevant, the level and type of aggregation (i.e. ways of spatially grouping values; an example is counting the population within different administrative zones). The level and type of aggregation are the subject of Section 4.9. This section discusses the issue of spatial scale. In this context spatial scale is defined as the scale at which the property of interest varies (but there are many definitions and Lloyd (2006) discusses some of these). For example, in a mountainous landscape, elevation values may differ a great deal between one location and another over very short distances. On a river flood plain elevation values may, in contrast, differ very little over even very large distances. In the former case, the spatial variation may be described as being of a fine scale or a high frequency. In the latter case, the spatial variation may be considered coarse scale or

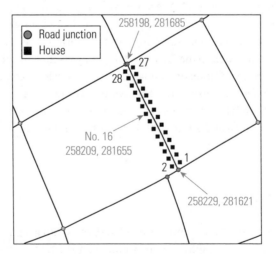

Figure 2.5 Geocoding: address interpolation.

low frequency. To capture spatial variation in the case of a mountainous terrain, a finer sampling grid would be needed than would be the case for the river flood plain. If the sample spacing is too large, then important information may be lost. If it is too small, then some of then effort expended in sampling will have been wasted—some of the data will be redundant and add little information. It is therefore important to consider how the available sample meets the requirements of the analysis and Section 2.8.1 discusses this issue further.

2.8 Spatial data collection

This section introduces the basic principles of some widely used means of spatial data collection. Spatial data may derive from secondary sources (e.g. paper maps) or data may be collected by, or for, a particular user with a particular purpose in mind (termed 'primary data'). The first subsection deals with the topic of spatial sampling. Paper-based secondary data sources are the focus of Section 2.6.2. Spatial data can be divided into those collected by remote sensing and those collected by ground survey, and Sections 2.6.3 and 2.6.4 introduce these two modes of data collection.

2.8.1 Spatial sampling

Any spatial data set is a sample. A remotely sensed image may cover the entire area of interest, but there are limits to the spatial resolution of such imagery. In any application making use of spatial data, it is necessary to find a balance between the amount of information required and the number and spatial positioning of observations. In terms of measurements on the ground, perhaps using GPS or some more traditional survey technology, making observations on a fine grid over the entire study area, will allow detailed characterization of the particular property of interest. However, such a strategy will be wasteful of effort and money if many neighbouring observations have similar characteristics, and thus contain very similar information (see the previous section for a related discussion). The objective of sampling design is, therefore, to design a sampling scheme whereby the maximum possible information is acquired for the minimum effort.

Various commonly used strategies for sampling exist. These may be based on making observations at locations with particular characteristics. Following the example of mapping elevations, making measurements at locations where there is a break of slope would be sensible. Other strategies are based on random selection of sampling locations, with a key objective being to minimize bias. Note that many routines exist for randomly selecting locations or values from a list. Using such approaches, a particular number of locations might be randomly selected from across the whole study area and measurements could then be made at these locations. Other strategies are based on, for example, random selection of locations within areas (a stratified sampling scheme), thus ensuring that particular areas are represented, but that the locations within them are determined randomly. One way of matching spatial variation to sample density is offered by the body of approaches known as geostatistics. With such an approach, the variogram (see Section 9.7) is used to characterize the spatial variation given some provisional survey and this information can be used to ascertain an optimal sample spacing.

Whether a user obtains the sample themselves or is reliant on data provided to them, it is necessary to be aware of the nature of the sample and to consider any potential issues that might arise in the use of the data as a function of the sample design. An introduction to spatial sampling is provided by Delmelle (2009).

2.8.2 Secondary data sources

In studies utilizing historic maps, or paper-based sources of which digital versions are not readily available, conversion from these paper-based sources to digital versions is necessary. Scanning provides a rapid means of deriving a raster image from paper maps, although it is rarely used now for spatial data generation. Where acquisition of information on particular features is desired then digitization of mapped features is often conducted. The most common approach is probably to scan the map and trace the edges of the features of interest to generate a new vector data layer; this is called 'heads-up' digitizing. Increasingly, spatial analyses are based exclusively on data that were collected directly in digital format and the often tedious process of tracing features from maps through digitizing is unnecessary for many users of GIS. An introduction to historical GIS, which includes discussion of conversion of historic paper maps into digital format, is the book by Gregory and Ell (2007).

2.8.3 Remote sensing

Huge amounts of spatial data are generated through various technologies that fall within the umbrella term 'remote sensing'. This may refer to airborne or spaceborne sensors (or simply cameras) or technologies for measuring subsurface characteristics such as electrical resistivity. Conventional aerial photography has been an important source of data historically. Figure 2.6 shows an example of an orthophoto in an area of Maryland in the USA. The data were obtained from http://terraserver-usa.com/ and more information about digital orthophotos is available at: http://online.wr.usgs.gov/ngpo/doq/. Orthophotos are remotely sensed images that have been georectified (see Section 2.5).

There are now many different kinds of airborne and spaceborne sensors that detect radiation from different parts of the electromagnetic spectrum. Sensors may be divided into two groups:

- passive sensors, which sense naturally available energy
- active sensors, which supply their own source of energy to illuminate selected features. Examples are radar and light detection and ranging (LiDAR). Radar emits pulses of microwave energy and LiDAR emits pulses of laser light.

Figure 2.6 Orthophoto of an area approximately 4 km west of Airedele, Maryland, USA. Area: 2.4 km by 1.6 km. Spatial resolution: 4 m. Image courtesy of the United States Geological Survey.

Sensors vary in terms of the spatial resolution (defined in Section 2.2.1) and the spectral resolution of the output imagery. The term 'spectral resolution' refers to which parts of the electromagnetic spectrum are measured. The number of available bands (i.e. parts of the spectrum) and the specific parts of the spectrum represented determine the uses to which the imagery can be put and these are likely to be selected with particular purposes in mind since different parts of the spectrum will highlight different characteristics of the Earth's surface. The majority of remote sensing systems collect information in one or more visible, infrared, or microwave parts of the spectrum (Lillesand *et al.*, 2007). The infrared parts of the spectrum are large compared to the visible part of the spectrum. The far infrared part of the spectrum, for example, is sensed in the acquisition of thermal imagery. Lillesand *et al.* (2007) provide a detailed account of remote sensing principles and practice.

Remote sensing is often used to derive topographic models, i.e. digital models of the surface of the Earth (or, indeed, elsewhere). The term 'digital elevation model' (DEM) refers to such a model and the most common form is a raster grid with cell values representing elevations above some arbitrary datum such as mean sea level. Some key technologies for constructing DEMs (including airborne LiDAR and radar-based systems) are summarized by Lloyd (2004). In the case of airborne LiDAR, outputs are 'point clouds' from which surfaces can be generated.

2.8.4 Ground survey

There are several widely used means of obtaining information on the spatial position of features through ground survey. Traditional survey techniques include tape and offset surveying and levelling. Ground survey techniques make use of the fact that we can measure the position of a point in three dimensions through the measurement of angles and distances from other positions. Angles may be measured using tools such as theodolites, while the advent of electronic distance measurers allows rapid and highly accurate measurement of distances. The measurement of angles and distances may be combined using a high-precision system called a total station. Total stations are capable of obtaining positional measurements accurate to 1 mm or so. Global positioning system (GPS) technology has a major role to play in ground survey. GPS receivers vary markedly in their characteristics and costs. They range from small hand-held systems capable of obtaining positional measurements accurate to within a few metres to differential systems costing tens of thousands of pounds (or dollars) but capable of obtaining positional measurements accurate to within 1 cm. With differential GPS, a stationary base station receiver is used to refine measurements made by one or more roving receivers. Another important technology for generation of spatial data is terrestrial LiDAR. This approach can be used to generate very accurate models of topographic surfaces and other objects on the surface of the Earth (see Pietro *et al.* (2008) for an example). Introductions to ground survey are provided by Lloyd (2004) and Longley *et al.* (2005a) and a very detailed account is the book by Bannister *et al.* (1998).

2.9 Sources of data error

All data are only representations of reality and are subject to a variety of factors which may affect their quality. This book outlines a wide range of methods for extracting information from data, but for these analyses to be worthwhile the data must be of sufficient quality. Sources of error in data include errors made during data collection, data input errors, and inappropriate data model choices. Identifying obvious errors caused by factors like equipment failure may often be straightforward, but many major sources of error may go unnoticed. Modelling and analysis of spatial data may introduce further errors. Manual conversion from paper-based to digital formats is a major potential source of error. In addition, any data collection technology has a limited precision even if that technology is employed properly and there are no external factors (such as atmospheric conditions) that have an impact on the accuracy of measurements. In terms of data model choice, if the spatial resolution of a raster image is coarse in relation to the area of objects of interest then those objects will not be well represented by the raster. A particular example is the representation of a linear feature like a river, which may be blocky in appearance if represented by a raster.

The accuracy of a data product can be conceived of as having two parts—bias and precision. A biased set of measurements may consistently over-estimate or under-estimate the 'true' value while precision refers to the repeatability of values. In other words, if there is apparently random variation in measurements and repeated measurements differ in some inconsistent way then the measurements are of low precision. In contrast, if repeated measurements are similar then the measurements are of high precision. The possibility of errors being introduced at any stage of data processing or analysis should be taken into account and the generation of high-quality graphic outputs should not disguise the fact that any output is only equal in quality to the lowest quality input. This topic is discussed further below. All spatial data should be associated with metadata—that is, data about the data sources which indicate key information on how and when the data were collected, as well as detailing any conversions or modifications undertaken. If detailed metadata are kept, these act as an invaluable resource for future users of the data in that they provide a means of assessing factors that may have an effect on applications which make use of these data.

2.9.1 Uncertainty in spatial data analysis

The use of alternative procedures, or selecting different options in the application of one method, will often lead to different results. It is essential in any use of spatial data to take into account such potential problems. The modelling of propagation of errors from one processing stage to another, and of the degree of uncertainty in representations of features and their attributes, are significant areas of research. The quality of outputs from a spatial analysis is a function of (1) the quality of the data, (2) the quality of the model, and (3) interactions between the data and the model (Burrough and McDonnell, 1998). When data from different sources are combined, the effects of many different kinds of uncertainties (e.g. measurement errors, scale differences, temporal differences, and other factors) may also combine. Spatial data quality and uncertainties in spatial data are among the subjects of the book chapter by Brown and Heuvelink (2008).

2.10 Visualizing spatial data

Visualization is the first stage of any spatial analysis. Simple viewing of a spatial data set may seem conceptually straightforward. However, there may be a multitude of decisions that have to be taken into account when visualizing data which may impact strongly on interpretations of those data and on the ways in which any analysis might proceed. Simple point patterns (i.e. point event locations with no attributes attached) are often presented using points to represent each event location. In the case of objects with categorical attributes (e.g. urban area or rural area), depiction may be based on the selection of different colours or shades to represent each category. In such cases, selection of colours or shades that enable differentiation between categories is important; a map with two classes depicted using similar shades or colours may be very difficult to use. Continuous variables (e.g. measurements of an airborne pollutant) are usually represented using a range of colours or shades (e.g. white for small values, shades of grey for intermediate values, and black for large values). Where the data model is a grid, a continuous grey scale or colour scale may be used, as shown in Figure 2.1.

A common means of displaying areal data (e.g. population densities in administrative zones) or values attached to other discrete objects such as points, in particular,

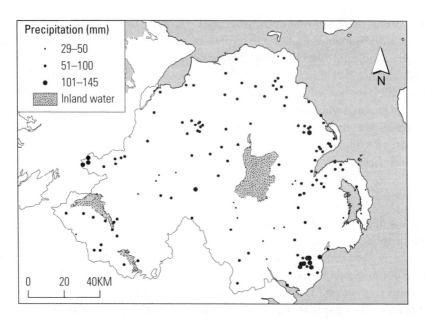

Figure 2.7 Precipitation amounts in Northern Ireland in July 2006.

is a choropleth map. With such maps, ranges of values are assigned a particular shading or colour. In other words, the possible values are divided into classes (typically five or less). Choropleth maps often show areas of uniformity separated from one another by abrupt edges (Tate *et al.*, 2008). The map is then a function of the classes used to display values and the form of the zones for which the values are provided—that is, in the same way that the zones are merely one possible way of spatially dividing a continuously varying phenomenon so are the classes used to represent values one way of subdividing the full range of values. It has been argued that one way of reducing such problems is to convert areal data into surfaces (Tate *et al.*, 2008; see Section 9.9). Figure 2.7 shows a map of point values (precipitation amounts) represented using symbols of different sizes; this is a common means of displaying point values. Even simple approaches open up complex issues—if a range of values is divided into five classes, the different class thresholds used may result in visually very different maps than those based on, say, six or seven classes. Figure 2.8 shows two different groupings of the same set of area values into three sets of different classes. While both maps are based on the same data, the patterns in the maps appear, in many respects, quite different.

There are many more sophisticated means of visualizing spatial data, including three dimensional visualizations (e.g. see Figure 9.5), familiar to users of Google Earth™ (http://earth.google.co.uk/), and cartograms. Cartograms distort the form of features to highlight particular characteristics, for example zones with large populations may be made proportionately larger than zones with small populations such that the modified zones better reflect the attributes that they contain. Throughout this book, maps and other visual outputs are presented as central components of the analyses of which they are part. Introductions to various aspects of spatial data visualization are provided

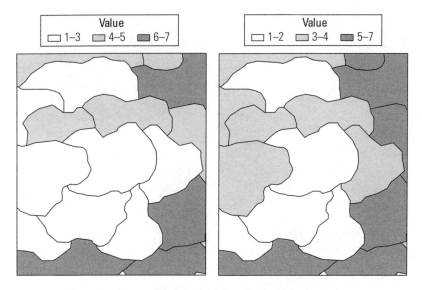

Figure 2.8 Alternative categorizations for vector polygon data.

in the book edited by Wilson and Fotheringham (2008). A good summary account of methods for visualizing spatial data (geovisualization) is given by Longley *et al.* (2005a). The data values may be visualized in other indirect ways, an example being the use of plots or graphs to summarize data values. This topic is amongst those explored in the following chapter.

2.11 Querying data

At its simplest level, the analysis of spatial data could involve selecting and mapping areas or features that have particular properties. For the example of measurement of some pollutant at point locations, all points with a pollution level above some critical threshold could be highlighted. In a standard database system a querying language (such as structured query language) is likely to be used to select entries in the tables that make up the database. Likewise, in a GIS such a query language can be used to select a subset of the data set. Such queries take logical forms such as Nitrogen-DioxidePPB > 21, selecting all locations where the nitrogen dioxide amount measured is greater than 21 parts per billion. The concept of Boolean logic and its application for querying spatial data is outlined next.

2.11.1 Boolean logic

Selection or combination of spatial features is often conducted using logical (or Boolean) operators (like the nitrogen dioxide example above). For example, population zones in

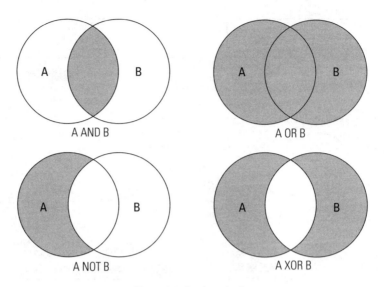

Figure 2.9 Boolean logic.

ID	Population
1	52
2	111
3	43
4	123
5	215
6	26
7	139
8	75
9	45
10	73
11	432
12	321
13	121
14	175
15	64
16	32
17	234
18	107

Selection for query:
Population < 100

Labels are ID numbers

Figure 2.10 Selected polygons and corresponding table entries.

a layer may be selected that have more than a specific number of people above a particular age, but less than a certain percentage (%) who have long-term illnesses. Such an argument would be structured like this: No.People75plus > 2500 AND LongTermIll% < 10. Figure 2.9 shows Venn diagrams (used to display possible combinations between groups) illustrating the outcome of the application of Boolean logic to two sets. In these cases, one, either, both, or mutually exclusive sets are selected. The Boolean AND selects objects fulfilling two criteria, OR selects objects fulfilling either of the criteria, NOT selects objects fulfilling one criterion but not the other and XOR (exclusive OR) selects objects fulfilling one criterion or the other, but not both. Figure 2.10 shows a simple example of a query in practice.

Queries constructed using Boolean logic can be used to select features in any desired combination. In the case of two or more criteria, query statements are easily extended. The application of Boolean logic for the overlay of multiple data layers (the identification of common areas of two or more sets of polygons) is discussed in Chapter 5. With Boolean logic, membership of a class is definite. An alternative approach is fuzzy logic, which recognizes uncertainty in assigning features to classes (see Longley *et al.* (2005a) for a summary). For example, boundaries between two soil types are not likely to be clearly defined and instead some form of classification that accounts for the probability of there being one soil type or another at a particular location is likely to be more appropriate than a 'hard' classification of the type described above (see Section 3.4 for a discussion about probabilities).

Summary

This chapter covers a wide variety of concepts that are important in the analysis of spatial data. The focus was on key GIS concepts, including data models, databases, projections, georeferencing and geocoding, spatial scale, spatial data collection, errors, visualization, and querying spatial data. Such issues are central to understanding the material covered in the rest of the book. Knowledge of data models is important as the data provide the basis of any analysis. Some understanding of database principles and data extraction (querying) is also central to a large proportion of analyses. Understanding of how data are collected, and the limitations of particular approaches, is essential background to the application of spatial data. No data are perfect representations of reality and so an awareness of potential sources of error is crucial. Finally, visualization and querying of spatial databases are common first steps in any spatial analysis.

Further reading

The further reading section of the previous chapter cited some useful introductions to GIS. Some of the books listed in that section provide in-depth material on some of the topics outlined in this chapter. In particular, issues such as databases, query of spatial data, and errors are dealt with by Burrough and McDonnell (1998), Longley *et al.* (2005a), and Heywood *et al.* (2006). Descriptions of spatial data formats and storage are given by Wise (2002). Spatial data collection is a vast topic; useful introductions to survey and remote sensing are provided by Bannister *et al.* (1998) and Lillesand *et al.* (2007), respectively.

3

Key concepts 2
Statistics

3.1 Introduction

This chapter is concerned with ways of summarizing and exploring numerical data. Even a brief summary of the key principles of statistics would require a dedicated book, so the intention of this chapter is to introduce some (very selective) ideas that it is necessary to understand to make use of parts of the rest of this book. These methods provide the basis of the *spatial* statistical methods that will be defined later on. The analysis of aspatial data (data with no spatial locational information) and spatial data usually starts with computation of standard summary statistics, as described in this chapter. Statistics can be divided into descriptive statistics, which provide summaries, and inferential statistics, which allow the making of inferences about a population (a complete data set representing all cases, e.g. all people in a country) from a sample. A sample is a partial data set, such as a population data set which excludes some people for some reason such as cost limitations, enabling only a limited survey. Both descriptive statistics and inferential statistics are introduced in this chapter, although more space is devoted to the former. Core concepts, which will be discussed in Section 3.4, include probabilities and the significance level. A statistical hypothesis may be associated with a probability that it is true or false and this is a central notion in statistics.

The following sections consider the purpose of statistical methods and introduce some ways of describing data sets. The focus here is initially on univariate statistics—methods that are used to analyse only one variable. Next, the focus is on multivariate methods—methods that deal with two or more variables simultaneously. In addition to introducing methods, the chapter will introduce some of the principles of statistical notation, for example one concern is to demonstrate how to 'read' the equations given in the rest of the book. Material of this nature is sometimes placed, for good reason, in the appendices of introductory books. In this book, such material is presented within the main text so that there is a direct transition from standard aspatial statistics (i.e. methods that do not take into account the spatial location of observations) to their spatial equivalents. This chapter deals exclusively with standard aspatial statistical methods. Methods which do take spatial location into account are the subject of later sections and Section 4.8 deals with the principles of one statistical approach to characterizing spatial variation (i.e. geographical patterning) in the property of interest.

Before proceeding, it is useful to consider how the kinds of data we have to work with may differ. Data may be divided into four types, which contain different amounts of information. These data types are:

Nominal   An arbitrary naming scheme, for example ethnic group (White, Caribbean, African).

Ordinal   Values are ordered, but there is no information on the relative magnitude of values, for example small, medium, large.

Interval   The intervals between measurements are meaningful, but there is no natural zero point, for example temperature. Differences between adjacent values are equal, i.e. 28–27 is the same as 99–98. Temperature (where zero is arbitrary, e.g. the freezing point of water for degrees Celsius) is often cited as an example of an interval variable (see, for example, Ebdon, 1985; O'Sullivan and Unwin, 2002). For two temperatures in degrees Celsius (e.g. 10°C and 20°C) and degrees Fahrenheit (50°F and 68°F), the ratios between the two sets of values are different: $^{10°C}/_{20°C}=0.5$ and $^{50°F}/_{68°F}=0.74$.

Ratio   Values with a natural zero point, ratios as well as intervals, are meaningful. For example, the ratio of 25 to 50 mm is the same as the ratio of the same measurements in inches (i.e. $^{25\,cm}/_{50\,cm}=0.5$ and $^{9.8\,in}/_{19.7\,in}=0.5$).

The main concern in this chapter is with the analysis of interval and ratio data and this is also the main focus in Chapters 8, 9, and 10.

3.2 Univariate statistics

The principal focus in this section is on what are termed 'descriptive statistics'—that is, methods to summarize or describe observations (measurements of some property). Summarizing an individual variable (e.g. precipitation amount) is done with reference to its distribution. The distribution of a variable refers to the set of values ordered from the smallest to the largest. Often, identical or similar values are grouped together, for example values 0–10 may be grouped, then values 11–20 and so on. In this way, we can refer to the frequency of values. For example, are most values very small with only a few large values or is there an even proportion of small and large values with most values being somewhere in between? When values are grouped into classes they can be depicted using a histogram. This is a form of chart that has bars of a size that is in proportion to the number of values in a given class. For example, if there are five observations in class 1 and 10 observations in class 2 then the bar representing class 2 will be twice as high as the bar representing class 1. Figure 3.1 shows a histogram with the range (minimum and maximum) of values represented by each bar indicated (e.g. values in the first bar range from 27.1 to 29). Frequency indicates the number of cases in a bin or class. As an example, there are 10 cases in the range 37.1 to 39.

Using too few or too many classes will not enable representation of important features of the distribution. The number of classes is likely to be determined as a function of the number of observations and the range of values that they take. If there are many thousands of observations (as might be the case, for example, using a remotely sensed image), and there is a sufficiently wide range of values, then it may be sensible to have a large number of classes, producing a 'smoother' distribution than would be possible for a smaller number of observations.

The most common way of summarizing a data set is to compute some kind of average, the mean average is the most well known. Averages are measures of central tendency in a distribution—in some sense the 'middle' value in the distribution. The mean average of the variable, and the notation used to represent this, is detailed below. First, a value of the variable is indicated by z_i. The value itself is given by z, and i is an index—it indicates the observation number. For example, if there are five observations

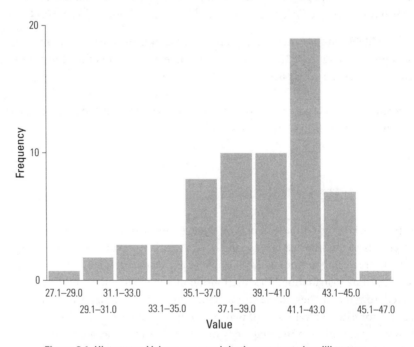

Figure 3.1 Histogram. Values are precipitation amounts in millimetres.

in the data set then i could take a value of 1, 2, 3, 4, or 5. The number of observations is indicated by n and, in the example given, $n=5$. The population (where population indicates it is assumed we have all possible values and not just a sample) mean average, μ (the Greek lower case letter mu), can be given by:

$$\mu = \frac{1}{n}\sum_{i=1}^{n} z_i \qquad (3.1)$$

The only term not yet explained is Σ (the Greek upper case letter sigma). Σ indicates summation. Below Σ is the term $i=1$ and above it is n. This means start at the first observation ($i=1$) and step through all values up to and including the last value ($i=n$). Note that other sources may use other letters for the variables, but the use of letters and symbols here is consistent throughout the text. $\sum_{i=1}^{n} z_i$ indicates that all values of z_i should be added together.

In the example above z_1 is taken first, then z_2 is added to it and so on until all values have been added together. The end result is then multiplied by $1/n$ (1 being the numerator (top part of the fraction) and n the denominator (bottom part of the fraction)). This gives the mean average of the values and is the same as dividing the summed values by n. As an example, if we have five values (z_1 to z_5) and they are 11, 14, 13, 9, and 6 then their sum is 53, $1/5=0.2$ and the mean average is given by $0.2 \times 53 = 10.6$.

Given these values, Equation 3.1 can be given as:

$$\mu = \frac{1}{5}\sum_{i=1}^{5} z_i = 0.2 \times (z_1 + z_2 + z_3 + z_4 + z_5) = 0.2 \times (11+14+13+9+6) = 0.2 \times 53 = 10.6$$

Other averages include the median (the middle value when all values are ordered from smallest to largest) and the mode (the most frequently occurring value). When there is an even number of values, the median is the mean average of the two values in the middle of the distribution (e.g. values 40 and 41 out of a total of 80 values with values ordered from smallest to largest). The mean average is very sensitive to outliers (i.e. unusually large or small values) and one benefit of using the median or mode is that the impact of outliers is reduced or non-existent.

The dispersion of a distribution is often of interest, i.e. how much variation is there in the values? The range—the absolute difference between the minimum and maximum value—is one simple measure. As noted above, the median is the middle of the ranked values. The value that falls 25% of the way along the list of ranked values (e.g. the mean average of values 20 and 21 of a total 80 values) is called the lower quartile and the value that falls 75% of the way along the ranked list (e.g. the mean average values of 60 and 61 of a total 80 values) is called the upper quartile. Together, the minimum, lower quartile, median, upper quartile, and maximum provide a summary of the distribution.

The dispersion around the mean—the degree to which values are close to the mean average—is given by the standard deviation. The standard deviation is small when the values are all quite similar to the mean and large when some values deviate markedly from the mean. The population standard deviation, indicated by σ (lower case sigma), is given by:

$$\sigma = \sqrt{\frac{1}{n}\sum_{i=1}^{n}(z_i - \mu)^2}$$

(3.2)

Most of the notation is familiar from the equation for the mean average. In this case, the mean (μ) is subtracted from each value and the product (i.e. the outcome) of this subtraction is squared. The squared differences are added together. Once this is done the sum is multiplied by $1/n$ and the square root is taken of this value. In words, the standard deviation is the square root of the average squared difference between observed values and their mean average. The squaring is necessary because if the difference between each value and its mean is not squared, then the sum of differences will be zero. Where the square root is not taken the resulting value is called the variance.

The mean and standard deviation as defined above are population statistics. In recognition of the fact that usually we have only a sample, an alternative form of the standard deviation is computed. The sample mean, $\bar{z}$ (z with a bar on top), is computed as above (i.e. the population mean in Equation 3.1). The sample standard deviation, s, is computed in the same way as in Equation 3.2 except that $1/n$ is replaced by $1/(n-1)$. The reason for this requires some explanation. The mean must be computed before we can compute the variance and a quantity known as the number of degrees of freedom is n minus the number of parameters (such as the mean) estimated, thus $n-1$ in this case (see O'Sullivan and Unwin (2002) for a further account). Another way of putting this is that one degree of freedom is used up in estimating the mean and if we know the mean then we only need $n-1$ values to calculate the value of the nth sample and thus know all values (i.e. if we have five values in total then we need only four values and the mean to work out the fifth value) (Rogerson, 2006). The sample standard deviation is given by:

$$s = \sqrt{\frac{1}{n-1}\sum_{i=1}^{n}(z_i - \bar{z})^2}$$

(3.3)

Note that population statistics are by convention given by Greek characters (e.g. σ) and sample statistics by ordinary lower case letters (e.g. s).

Using the same data as before (11, 14, 13, 9, and 6 with a mean average of 10.6), the sample standard deviation is calculated using:

$$\sum_{i=1}^{5}(z_i - \bar{z})^2 = (11-10.6)^2 + (14-10.6)^2 + (13-10.6)^2 + (9-10.6)^2 + (6-10.6)^2$$

$$= 0.16 + 11.56 + 5.76 + 2.56 + 21.16 = 41.20$$

Given this, s is obtained:

$$s = \sqrt{\frac{1}{5-1} \times 41.20} = \sqrt{0.25 \times 41.20} = 3.21$$

The mean and standard deviation are useful measures of a distribution if it is normal. A normal distribution is characterized by an equal proportion of small and large values, with a peak of values in the middle ranges—the distribution is symmetric. This type of distribution is called bell-shaped. A distribution with a large number of small values and a small number of large values is termed 'positively skewed', while a distribution with a small number of small values and a large number of large values is termed 'negatively skewed' (an example is the histogram in Figure 3.1). The mean average is 'pulled' in the direction of the skew, i.e. it is affected by extreme values. In a skewed distribution, the mode will be under the peak of values, the mean will be closer to the 'tail' of extreme values, and the median will be in between the mode and the mean. The degree of skewness can be measured by a statistic called the coefficient of skewness (note that different measures of skewness exist; the measure below is as implemented in Microsoft® Excel®). This can be given by:

$$\text{skewness} = \frac{n}{(n-1)(n-2)} \sum_{i=1}^{n} \left(\frac{z_i - \overline{z}}{s} \right)^3 \tag{3.4}$$

In words, the right-hand side of Equation 3.4 is the sum of the cubed product of differences between individual values and their mean average divided by the standard deviation. Positive values of the skewness coefficient indicate positive skew and negative values indicate negative skew, while a value of zero indicates no skew. Some examples are given in Figure 3.2.

Some distributions have two or more modes, i.e. peaks of values. It is important to examine the distribution of a variable prior to further analysis. Examples of distributions which are normal, positively skewed, and negatively skewed are given in Figure 3.2; for the purposes of the discussion the data are treated as measurements of precipitation amount in millimetres. If a distribution is normal (and we can perceive it as a bell-shaped smooth curve, rather than a histogram with discrete bars, as shown in Figure 3.1) then 68.26% of the values in the data set should fall within one standard deviation of the mean. In other words, 68.26% of the area under the normal curve lies within one standard deviation of the mean (i.e. above or below the mean), 95.46% of the area lies within two standard deviations, and 99.73% lies within three standard deviations. If the distribution is normal, the mean is 10.6, and the standard deviation is 3.21 then 68.26% of the values should be 10.6±3.21 (i.e. in the range 7.39 to 13.81).

Following the discussion above, the mean and standard deviation are not representative if the distribution is not close to normal—this potentially affects many of the procedures detailed in Chapters 8, 9, and 10 in particular. Various possible solutions exist: the variables can be transformed (e.g. by taking the square root or the log of the values) and the transformed variables may have a less skewed distribution. Analysis can

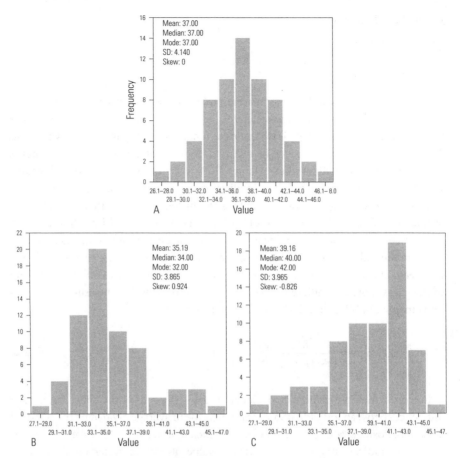

Figure 3.2 Histograms: (A) normal distribution, (B) positive skew, (C) negative skew. Values are precipitation amounts in millimetres. SD is the standard deviation and Skew is the skewness coefficient.

then be conducted using the transformed variables and these can be back-transformed (i.e. converted back to the original values) after the analysis. The logarithmic transformation is used widely to account for distributions with long tails of positive values (i.e. a small proportion of large values); in such cases, the distribution of the log-transformed data should be approximately normal and the transformed data can then be analysed in the usual way. A logarithm of a number is that particular number expressed as a power of another number. Natural logarithms are numbers expressed as powers of *e* (often given by exp), which is the exponential constant (approximately equal to (this term is indicated by ≈) 2.718281829; see Appendix B for more details) while common logarithms are expressed as powers of 10 (Shennan, 1997). For example, 42 to the natural base ($\log_e$) is 3.738, which can be given by $e^{3.738}$, and 42 to the base 10 ($\log_{10}$) is 1.623, which can be given by $10^{1.623}$. Logarithms can be calculated using standard spreadsheet and GIS packages. Introductions to transformations are provided by Gregory (1968) and Shennan (1997).

The next section is concerned with analysing two or more variables simultaneously.

3.3 Multivariate statistics

The focus so far has been on methods for exploring single variables. In many applications, there is a need to consider how two or more variables are related to one another. For example, what is the relationship between altitude and snowfall? Does snowfall tend to be greater at high altitudes? This section makes use of the data given in Table 3.1. The first stage of an analysis may involve plotting one set of values against the other (this is called a scatter plot—an example is given in Figure 3.3) and such an analysis could be expanded using regression analysis, as detailed in this section.

Figure 3.3 indicates that small values of z (in this example, snowfall) tend to correspond to small values of y (elevation) and that large values of one tend to correspond

Table 3.1 Sample data set for illustrating regression.

Variable 1 (y) in m	Variable 2 (z) in cm
12	6
34	52
32	41
12	25
11	22
14	9
56	43
75	67
43	32

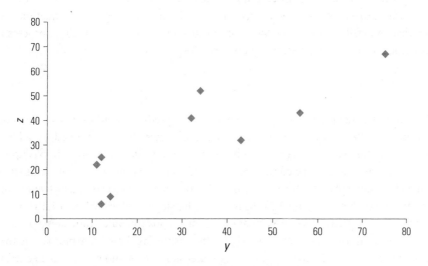

Figure 3.3 Scatter plot. Elevation (y) in metres (m) against snowfall (z) in centimetres (cm).

to large values of the other. Put another way, in general an increase in one corresponds to an increase in the other—it can be said that the variables are positively related to one another. If, as the values in one variable increase, the values of the other variable decrease (or vice versa) then the relationship is said to be negative. Analysis of relationships often proceeds using correlation and regression, which enable exploration of the nature of the relationship between two or more variables and the strength of the relationship between them. Correlation and regression are explored in this section.

In the case of two variables, regression is used to fit a line through the points on a scatter plot, this line being as close as possible to all the points according to some criterion. This is called the line of best fit. The line represents the trend in the data. If the variables are positively related, the line will be low with respect to the z-axis (representing small value) on the left of the graph and will increase diagonally from left to right. This would be the case for a line fitted to the plot in Figure 3.3. The correlation coefficient, r, provides a measure of the nature and strength of the relationship between variables. More specifically, it can be interpreted as indicating the degree to which points scatter around the regression line. Before detailing the measurement of correlation, the procedure for fitting a line to the scatter plot is detailed.

In this example, the variable y (elevation) is the independent variable while the variable z (snowfall) is the dependent variable. As well as allowing for exploration of the nature of the relationship between two variables, regression enables prediction of the values of dependent variables given values of independent variables. For example, if we have a raster map of elevation values across a region (i.e. we have values at all locations of interest) but only a few snowfall measurements, we could conduct regression by taking elevation values at the snowfall measurement locations and plotting these against the snowfall values. Once a line is fitted, the regression equation (indicating the form of the fitted line) can be used to predict snowfall values at locations where there are no snowfall measurements because the regression equation tells us what snowfall amounts to expect for any given value of elevation. This process is described below. The regression equation can be given by:

$$\hat{z}_i = \beta_0 + \beta_1 y_i \qquad (3.5)$$

This indicates that the predicted value of z_i (with a prediction indicated by the hat on top of the letter)—the value given by the line of best fit—is obtained by adding β_0 to β_1 multiplied by y_i (in this example, the elevation value at location i). β is upper case Greek beta and these components are referred to as the beta coefficients. β_0 is called the intercept and is the point where the line crosses the vertical axis (representing the z variable in Figure 3.3). β_1 is called the slope coefficient. A negative value for the slope indicates a negative relationship and a positive value indicates a positive relationship. In this case, we know β_1 will be positive as the scatter plot shows that a higher elevation will correspond to a greater amount of snow. What is needed is a method to identify appropriate values of β_0 and β_1. Once we have these, we have figures that tell us something about the nature of the relationship between variables. The least squares method is the most common approach to fitting a line to data and obtaining values of β_0 and β_1. This method minimizes the squared difference between the observed value (z_i, the measurement) and the value given by the line fitted with regression ($\hat{z}_i$):

$$\sum_{i=1}^{n}(z_i - \hat{z}_i)^2 \tag{3.6}$$

The following text describes how the intercept and slope are obtained through the ordinary least squares (OLS) method. The slope coefficient, β_1, is obtained from:

$$\beta_1 = \frac{\sum_{i=1}^{n}(y_i - \overline{y})(z_i - \overline{z})}{\sum_{i=1}^{n}(y_i - \overline{y})^2} \tag{3.7}$$

The numerator gives the covariance between the independent and dependent values. The covariance is a measure of the degree to which two variables vary together and is the difference in one value from its mean multiplied by the difference in the second value from its mean. The covariances for each location are summed. The denominator is the sum of squared differences between the independent values and their mean.

The intercept, β_0, is given by:

$$\beta_0 = \frac{\sum_{i=1}^{n}z_i - \beta_1\sum_{i=1}^{n}y_i}{n} = \overline{z} - \beta_1\overline{y} \tag{3.8}$$

The values used in Table 3.1 are used to illustrate the regression procedure. Note that this sample is very small and in practice regression analyses should be based on much larger samples. However, this small sample allows direct illustration of the methods. This topic is discussed further in Section 3.4.

The slope is computed first. Initially, we compute the numerator of Equation 3.7:

$$\sum_{i=1}^{n}(y_i - \overline{y})(z_i - \overline{z})$$

We take each value of y and subtract the mean value of y; in turn we take each value of z and subtract the mean value of z. The difference between each y value and its mean and each z value and its mean is then multiplied together as shown in Table 3.2 (in the column headed $(y_i - \overline{y}) \times (z_i - \overline{z})$). This is done for all of the observations and the multiplied values are added together. The mean value of y is 32.11 and the mean value of z is 33.

The sum of the multiplied differences in Table 3.2 is 3092. The denominator of Equation 3.7, $\sum_{i=1}^{n}(y_i - \overline{y})^2$, is then used. In words, we take each value of y, subtract its mean, square this difference (see the column headed $(y_i - \overline{y})^2$ in Table 3.2), and add

Table 3.2 Variable 1 (y) and variable 2 (z), differences from their mean, differences multiplied, and the square of the differences from the mean of variable 1.

Variable 1 (y_i)	Variable 2 (z_i)	$(y_i - \bar{y})$	$(z_i - \bar{z})$	$(y_i - \bar{y}) \times (z_i - \bar{z})$	$(y_i - \bar{y})^2$
12	6	−20.11	−27.00	543.00	404.46
34	52	1.89	19.00	35.89	3.57
32	41	−0.11	8.00	−0.89	0.01
12	25	−20.11	−8.00	160.89	404.46
11	22	−21.11	−11.00	232.22	445.68
14	9	−18.11	−24.00	434.67	328.01
56	43	23.89	10.00	238.89	570.68
75	67	42.89	34.00	1458.22	1839.46
43	32	10.89	−1.00	−10.89	118.57

all of these squared differences together. The sum of these squared differences is 4114.89.

To compute the slope value, β_1, we divide the first summed value by the second:

$$\frac{3092}{4114.89} = 0.75142$$

The intercept, β_0, is then calculated:

$$\bar{z} - \beta_1 \bar{y} = 33 - (0.75142 \times 32.11) = 8.87190$$

In words, the intercept is given by the mean of the z values minus β_1 multiplied by the mean of the y values ($\beta_1 \bar{y}$ means that the two components are multiplied by one another and no multiplication symbol is needed). Note that a fairly large number of decimal places are used in the calculations to ensure that the manual calculations are close to those obtained using software packages.

The fitted line is shown in Figure 3.4. Regression is more conveniently conducted (i.e. values for β_0 and β_1 obtained) using matrix algebra, as outlined below, and such an approach is used in computer algorithms. The plot was generated using a spreadsheet package. Note that the intercept value in Figure 3.4 is slightly different to the figure given above, and the difference is due to rounding error in the calculations. The r^2 term in Figure 3.4 is the coefficient of determination and it is defined below.

Once we have values of β_0 (the intercept) and β_1 (the slope coefficient), we can make predictions. As an example, using the regression line shown in Figure 3.4, suppose we have a location with no snowfall measurement, but we know the elevation at that location is 43 units (e.g. metres). Following the regression equation

$$\hat{z}_i = \beta_0 + \beta_1 y_i$$

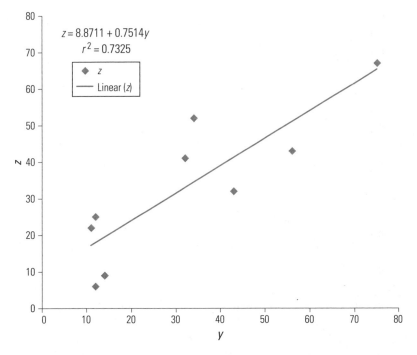

Figure 3.4 Scatter plot. Elevation (y) in metres (m) against snowfall (z) in centimetres (cm) with line of best fit.

we replace β_0 and β_1 with the values obtained for these previously and replace y_i (an elevation value in this case) with 43. This leads to:

$$\hat{z}_i = 8.87190 + (0.75142 \times 43) = 41.18296 \text{ cm}$$

For an elevation value of 43, therefore, the predicted value of snowfall (to three decimal places) is 41.183. This can be confirmed by looking at Figure 3.4 and drawing a line from the point corresponding to approximately 43 on the y (elevation) axis upwards to meet the line of best fit and then drawing a line from the point where the added line and the line of best fit meet across to the z (snowfall) axis. If the lines are accurately drawn, then a value of approximately 41 can be identified on the z (snowfall) axis.

The goodness of fit of a line of best fit can obtained by measuring the residuals, i.e. the difference between observed values and predicted values. As an example, Table 3.1 includes a y (elevation) value of 11 paired with a z (snowfall) value of 22. Using the approach outlined, the predicted value of snowfall for an elevation value of 11 is given by:

$$\hat{z}_i = 8.87190 + (0.75142 \times 11) = 17.13752 \text{ cm}$$

In this case the observed value is 22, and the predicted value to three decimal places is 17.138 so there is a difference (residual) of −4.862. In words, the regression model

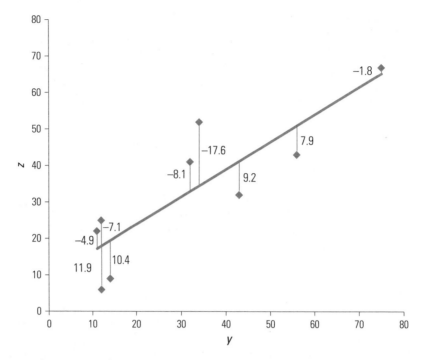

Figure 3.5   Scatter plot. Elevation (y) in metres (m) against snowfall (z) in centimetres (cm) showing residuals.

under-predicts the observed value by 4.862. Figure 3.5 shows residuals (to one decimal place) for the scatter plot in Figure 3.4.

Examining the residuals, perhaps by mapping them, may prove illuminating and may help to highlight regions with unusual characteristics. One goal of regression is to minimize the squared residuals while another is to minimize clustering in the values of the mapped residuals (see Sections 4.8 and 8.2 for discussions on a related topic).

Before discussing measures of goodness of fit, some grounding is briefly given in more efficient means for obtaining regression coefficients than was provided above. This background is necessary to enable readers to make the most of the descriptions of local regression procedures that come later. The following text introduces the idea of matrices and matrix multiplication—concepts essential to the worked example. Another key topic, inversion, is outlined in Appendix E.

Using matrix notation, the ordinary least squares (OLS) regression (the standard method of finding regression coefficients as outlined above) coefficients can be obtained using:

$$\beta = (\mathbf{Y}^T\mathbf{Y})^{-1}\mathbf{Y}^T\mathbf{z} \qquad\qquad (3.9)$$

The upper case bold letters indicate a set of values that can be arranged in a rectangle with at least two rows and columns. The lower case bold letter indicates the case with only one column; this latter type of matrix is called a vector. The following are examples for the case of two variables and five cases (observations) of each:

$$\mathbf{Y} = \begin{bmatrix} 1 & y_1 \\ 1 & y_2 \\ 1 & y_3 \\ 1 & y_4 \\ 1 & y_5 \end{bmatrix}, \quad \mathbf{Y}^T = \begin{bmatrix} 1 & 1 & 1 & 1 & 1 \\ y_1 & y_2 & y_3 & y_4 & y_5 \end{bmatrix}, \quad \mathbf{z} = \begin{bmatrix} z_1 \\ z_2 \\ z_3 \\ z_4 \\ z_5 \end{bmatrix}$$

$\mathbf{Y}^T$ indicates the transpose of $\mathbf{Y}$—the flipped version of the original matrix (values in the left-hand column of $\mathbf{Y}$ are the top row in $\mathbf{Y}^T$ and values in the right-hand column of $\mathbf{Y}$ are the bottom row in $\mathbf{Y}^T$). The superscript −1 indicates the inverse, and this is explained in Appendix E. The '1' values for each entry in $\mathbf{Y}$ indicate that we are fitting a constant (intercept). In this example, we have five 1s, and five values of each of the independent (y) and the dependent (z) variables. Given Equation 3.9 and a knowledge of matrix algebra it is possible to find the regression coefficients for any number of independent variables. Computers make use of matrices, and so some understanding of how to use such methods is useful. Appendix E shows how Equation 3.9 is solved.

The solution to Equation 3.9 for the data presented above is given by (with the full working given in Appendix E):

$$\beta = (\mathbf{Y}^T\mathbf{Y})^{-1}\mathbf{Y}^T\mathbf{z} = \begin{bmatrix} 0.36169 & -0.00780 \\ -0.00780 & 0.00024 \end{bmatrix} \times \begin{bmatrix} 297 \\ 12629 \end{bmatrix} = \begin{bmatrix} 8.871 \\ 0.751 \end{bmatrix}$$

where the calculations used to obtain the final values (the intercept and slope), shown in the matrix to the right-hand side above, are:

$$\beta_0 = (0.362 \times 297) + (-0.0078 \times 12629) = 8.871 \quad \text{(intercept)}$$
$$\beta_1 = (-0.0078 \times 297) + (0.00024 \times 12629) = 0.751 \quad \text{(slope)}$$

Note that the value of β_0 is smaller than the value obtained previously. This is due to rounding error (i.e. a different number of decimal places used in calculations).

The strength of the relationship between the variables can be measured using the correlation coefficient. The correlation coefficient, r, is given by:

$$r = \frac{\sum_{i=1}^{n}(y_i - \bar{y})(z_i - \bar{z})}{\sqrt{\sum_{i=1}^{n}(y_i - \bar{y})^2}\sqrt{\sum_{i=1}^{n}(z_i - \bar{z})^2}} \tag{3.10}$$

The numerator is the same as for the estimation of β_1 (see Equation 3.7). The denominator is simply the square root of the sum of squared differences between y

Table 3.3 Variable 1 (y) and variable 2 (z), squared differences from their mean, and summed values.

Variable 1 (y_i)	Variable 2 (z_i)	$(y_i - \bar{y})^2$	$(z_i - \bar{z})^2$
12	6	404.46	729.00
34	52	3.57	361.00
32	41	0.01	64.00
12	25	404.46	64.00
11	22	445.68	121.00
14	9	328.01	576.00
56	43	570.68	100.00
75	67	1839.46	1156.00
43	32	118.57	1.00
	Sum	4114.89	3172.00

and its mean multiplied by the square root of the sum of squared differences between z and its mean. These two sets of products are then multiplied together.

We have already seen the differences between y and z and their respective means in Table 3.2. The squared differences and their sums are given in Table 3.3 (note that the values in column 3 are the same as those in the final column of Table 3.2).

Given the values in Table 3.3, the numerator is obtained from:

$$\sum_{i=1}^{n}(y_i - \bar{y})(z_i - \bar{z}) = 3092 \quad \text{(as calculated above)}$$

and the denominator is obtained from:

$$\sqrt{\sum_{i=1}^{n}(y_i - \bar{y})^2}\sqrt{\sum_{i=1}^{n}(z_i - \bar{z})^2} = \sqrt{4114.89} \times \sqrt{3172.00}$$
$$= 64.1474 \times 56.3205$$
$$= 3612.8136$$

The correlation coefficient is then given by:

$$r = \frac{3092}{3612.8136} = 0.8558$$

The r value is positive and indicates, as the slope value, that the variables are positively related: as the value in one variable increases so does the value of the other. If the r value was negative, this would indicate negative correlation: as the value in one variable decreased, the value of the other would increase. Possible values of r range from −1 (indicating perfect negative correlation) to +1 (indicating perfect positive correlation).

The coefficient of determination is often used to indicate the goodness of fit, which is simply the squared correlation coefficient and is given by r^2. In our example (given three decimal places), $r^2 = 0.856^2 = 0.732$. In this case, the coefficient of determination indicates that 73.2% of the variation in the data can be explained by the line of best fit. In words, the model represents the relationship quite well. So, an r^2 value close to zero indicates that the line (model) is a poor fit, while an r^2 value close to one indicates that the model is a good fit. Of course, where the relationship is non-linear (e.g. the scatter plot shows, along the horizontal axis, large values followed by small values, followed by large values, in a 'V' shape) then r (and r^2) may be close to zero and the scatter plot plays a key role in interpretation (Rogerson, 2006). The regression and correlation examples in this section are based on two variables (the dependent and an independent variable). Regression can easily be expanded to include more than one independent variable, thus allowing the assessment of the interrelationships between several variables simultaneously. In the case of more than one independent variable, upper case characters are used for the correlation coefficient and the coefficient of determination, thus R and R^2.

Some forms of data (e.g. nominal or categorical variables) should not be analysed directly using the methods outlined above. Alternative approaches are available in the case of values that are constrained to be whole numbers. Percentages and proportions should first be transformed before their analysis using standard statistical methods; Aitchison (1986) details some appropriate methods.

The topic of the following section is inferential statistics (i.e. statistical methods for making inferences about a population from a sample as opposed to descriptive statistics, which have been the focus in this section) and significance testing (e.g. testing for the significance of the differences between groups). As an example, it is standard practice to ascertain the significance of regression coefficients or the correlation coefficient, and the testing of the latter is outlined below.

3.4 Inferential statistics

In this section, the focus is on statistical methods for making inferences about a population from a sample as opposed to statistics which simply summarize a sample. Two common tasks in inferential statistics contexts are (1) to consider the likelihood that a statement about a given parameter (e.g. the mean or standard deviation) is true given the available data and (2) to estimate the parameters (Brunsdon, 2008). The first of these relates to the concept of hypothesis testing while in the second the confidence interval is central.

A common objective in statistical inference is to compare sets of samples and assess the degree of difference between the samples. In other words, we may be interested in assessing the probability that two samples come from different populations. Comparison of samples is based on tests of significance. In words, we test the significance of the difference between two (or more) samples to assess if the difference between them is likely to be 'real' in some sense. Questions of differences between samples are usually phrased in terms of a null hypothesis and the alternative hypothesis, indicated by H_0 and H_1, respectively. In simple terms, H_0 could be the hypothesis that there is no significant difference and H_1 the hypothesis that there is a significant difference. The significance level, α (lower case Greek alpha), can be defined as the probability (defined below) that the null hypothesis is correct (Ebdon, 1985). If some outcome is statistically significant then it is regarded as unlikely to have occurred by chance. Hypothesis testing is therefore the procedure of rejecting or accepting the null hypothesis. The possible errors associated with this process can be defined as follows:

type I error: reject the null hypothesis when it is true

type II error: accept the null hypothesis when it is false

The emphasis is on avoiding type I errors on the grounds that accepting the null hypothesis when it is false is likely to be less damaging than wrongly accepting the hypothesis that there are real differences.

It is useful to have some indication of how likely it is that a sample is representative of a population. For example, if we have the mean of the salaries of a set of individuals how confident can we be that this is representative of the population as a whole in the study area? This question can be approached by computing the standard error of the mean:

$$SE_\mu = \frac{s}{\sqrt{n}} \tag{3.11}$$

where μ is the mean average, s is the sample standard deviation, both as defined previously, and n is the number of observations in the sample. The standard error determines what is known as the confidence interval and a small standard error gives greater confidence that the sample mean is close to the population mean. The next stage of the discussion requires some knowledge of the concept of probability. A probability can be defined as the likelihood of a given outcome and is expressed as a fraction of 1. If a given event occurs 70 times out of 100 then the probability of the event occurring is expressed as 0.7. Given Equation 3.11, for a normal distribution, there is a 0.682 probability that the population mean is within one standard error of the mean and a 0.954 probability that the population mean is within two standard errors of the mean (see Section 3.2):

$$SE_\mu = \frac{4.14}{\sqrt{64}} = \frac{4.14}{8} = 0.518$$

In words, there is a 0.682 probability that the population mean is 37 ± 0.518 (or 36.482 to 37.518). By widening the confidence interval, we can say that there is a 0.954 probability that the population mean is 37 ± 1.035 (or 35.965 to 38.035).

As well as assessing confidence in the mean of a single sample, there is often a desire to assess differences between two different sample means. One way of doing this is to use the t-statistic (introduced in a specific context below). Introductions are provided by Shennan (1997), Ebdon (1985), and Rogerson (2006). When the desire is to compare multiple categories, analysis of variance (ANOVA) represents the standard framework. The ANOVA test can be used to compare variation within data columns to variation between data columns. In words, given a null hypothesis that a set of population means are equal, we would reject the null hypothesis if the variation between the means of each group is significantly greater than the variation within the data columns (Rogerson, 2006).

Section 3.3 discussed correlation and regression. If it is assumed that the distributions of the variables are normal and the observations of each variable are independent of one another, then a test of the significance for the correlation coefficient, r, may be conduced using the t-statistic (Rogerson, 2006):

$$t = \frac{r\sqrt{n-2}}{\sqrt{1-r^2}} \qquad (3.12)$$

Given the small example in Section 3.3, where r was 0.856 and n was 9, this gives:

$$t = \frac{0.856\sqrt{9-2}}{\sqrt{1-0.856^2}} = \frac{0.856\sqrt{9-2}}{\sqrt{1-0.856^2}} = \frac{2.265}{0.517} = 4.381$$

Before assessing this result, a little explanatory text is required. Hypotheses can be one-sided or two-sided. In the former case, a test is conducted to assess if the true value is above *or* below (but not both) a given value. In the latter case, the test considers if the true value is *either* side of a given value. This example relates to a two-sided (or two-tailed) test.

Given these values, examining a t-table (a table allowing identification of the significance level associated with a t value given a particular number of degrees of freedom (see, for example, Shennan, 1997; Ebdon, 1985)) indicates that, for a two-tailed test with $\alpha = 0.05$ (i.e. the 5% significance level, commonly used as a benchmark significance level) and with seven degrees of freedom (there are nine observations and $n-2=9-2=7$) the critical value of t is 2.365. Since 4.381 is greater than that value, the correlation coefficient can be said to be not equal to zero and the null hypothesis is rejected. An alternative is to use one of the many web-based t-test calculators. Of course, standard software packages will do the calculations for you in any case. It is important to take into account the sample size when assessing the correlation coefficient (or other coefficient) as, while the correlation coefficient may suggest a strong relationship between variables, there may in fact be little confidence in the results if the sample size is small.

3.5 Statistics and spatial data

The focus of this book is the analysis of *spatial* data. Spatial data cannot be blindly treated in the same way as data that are not located spatially (i.e. aspatial data).

One key problem relates to statistical inference, as discussed above. One of the assumptions of standard significance tests is independence in the observations of the samples. In simple terms, this means that the value of one observation should not be affected by the value of another observation. As will be shown in Section 4.8, values of observations of spatial variables are often very similar to the values at neighbouring locations. Indeed, the fact that this is so frequently the case provides the basis of many spatial data analysis approaches. In the present context, the problem with this is that significance levels tend to be inflated as the effective number of degrees of freedom is reduced—that is, if the observations are spatially dependent then there are effectively fewer independent observations. In words, use of standard significance tests with spatial data is problematic and solutions may not be straightforward. Rogerson (2006) provides a detailed discussion of this topic.

Summary

This chapter provides a summary of some key ideas and methods in descriptive and inferential statistical analysis. Many methods for the analysis of spatial data have their origins in standard aspatial statistical methods, and knowledge of some key methods and issues in statistical analysis is essential for the developing spatial analyst, who will need to make frequent use of standard statistical approaches in the initial exploration of spatial data sets. Many of the themes introduced here will be revisited in later chapters, but with adaptations reflecting the focus of this book on spatial data.

Further reading

The book by Rowntree (2000) provides an excellent introduction to some important statistical concepts. Various books provide introductions to statistics for geographers and the book by Rogerson (2006) is a good example. The books by Kitchin and Tate (2000) and O'Sullivan and Unwin (2002) also include relevant introductory material. Rowntree (2000) and Rogerson (2006) include very clear introductions to some key concepts in inferential statistics. Brunsdon (2008) provides an introduction to some key concepts in statistical inference and discusses some issues related to inference in a spatial context. Section 7.3.1 illustrates the chi-square (χ^2) test, a commonly used statistical hypothesis test.

→ The next chapter explores some key concepts in *spatial* data analysis and is the third of the three chapters introducing key concepts that provide foundations for the rest of the book.

Key concepts 3
Spatial data analysis

4.1 Introduction

In the previous chapter, a key focus was on introducing some methods for aspatial data analysis. This chapter builds on that previous discussion and discusses some ways of extracting information from spatially referenced (mappable) data. The chapter details how basic measurements, including lengths, perimeters, and areas, are made in a GIS. Following this, the generation of buffers (distance bands around specified objects) is detailed. Approaches for use with both vector and raster representations of features are considered. Next, the idea of moving windows—whereby some operation (such as computing the mean average) is conducted using local subsets of the data—is introduced. The subject of geographical weights, another core component of many methods detailed in the book, is outlined next. With such approaches, the tendency for nearby values to be more similar than those that are more distant is taken into account. A section on spatial dependence and spatial autocorrelation discusses spatial patterns and some ways of analysing such patterns. The basis of spatial dependence is that values close together in space tend to be more similar than those that are farther apart. This principle has been referred to as the 'first law of geography' (the concept is outlined by Tobler, 1970) and this concept is central to many methods for the analysis of spatial data. The need to consider spatial scale and the form of zones (where used) in any analysis is discussed in the following section. A small section on merging polygons then follows. Finally, the key themes are revisited and summarized.

In short, the objective of the chapter is to introduce the basic components of some key tools for the analysis of spatial data. Once these ideas (and those presented in the previous two chapters) are grasped, all of the background necessary to understand (at a simple level) the rest of the material presented in the book will have been developed. The key components of the chapter can be summarized as follows:

- Measuring distances.
- Measuring lengths and perimeters.
- Buffers—measuring zones of fixed distances around objects.
- Moving windows—mapping how values change from place to place. Moving windows are used in many contexts, including ascertaining the gradient or aspect of the terrain locally.
- Geographical weighting—for a given location, giving more influence to close-by values (e.g. estimating the mean at a given location, but giving greater weight to close-by values than to values further away).
- Spatial dependence and spatial autocorrelation—measuring the degree of similarity in neighbouring values (or values separated by a particular distance).
- The ecological fallacy and the modifiable areal unit problem—making inferences from aggregated data (e.g. numbers of people aged over 65 in an area) and considering changes in results due to changes in the size and shape of zones (e.g. using large administrative zones or smaller zones that fit within them).
- Merging polygons—joining subregions to form new larger regions.

4.2 Distances

Much of spatial data analysis relies on measuring straight line (Euclidean) distances between different locations. The distance, d, between point i and point j is calculated using Pythagoras' theorem:

$$d_{ij} = \sqrt{(x_i - x_j)^2 + (y_i - y_j)^2} \tag{4.1}$$

where location i has the coordinates x_i, y_i and location j has the coordinates x_j, y_j. In words, the squared difference between the two x coordinates and the squared difference between the two y coordinates are calculated and added together, and the square root of the product is taken. As an example, take a location with an x coordinate of 10 and a y coordinate of 15, and a second location with an x coordinate of 22 and a y coordinate of 19. The distance between these two locations is obtained from:

$$\sqrt{(10-22)^2 + (15-19)^2} = \sqrt{144+16} = \sqrt{160} = 12.649$$

In some cases, straight line distances may not be meaningful. As well as Euclidean distances, Manhattan distances (also referred to as taxicab distances) are also widely used. These refer to distances along grids and the name derives from the grid-like configuration of streets in Manhattan, New York. Manhattan distances are the sum of distances along each grid segment connecting the start and end points. Network distances are distances along networks. An example is distances along a road network, using the kind of vector structures detailed in Section 2.2. This book details a variety of ways of representing distances. These include friction surfaces, whereby the 'cost' of moving over a particular area of land is taken into account (see Section 10.6).

4.3 Measuring lengths and perimeters

With raster grids, lengths can be measured along cells (as outlined above) or in terms of Euclidean distances between, for example, cell centroids. In the former case, measurement requires information on the spatial resolution of cells and their number. In a simple case, if we are measuring the length along the side of five cells and their spatial resolution is 10 m, then clearly the distance is $5 \times 10 = 50$ m. Another way of dealing with distance travelled over raster grids, the use of friction (cost) surfaces, is discussed in Section 10.6. Measurement of lengths of vector features is discussed next.

4.3.1 Length of vector features

Lines can be measured simply by calculating the length of each line segment using Pythagoras' theorem (see Section 4.2) and summing the lengths of each segment that makes up a line. Perimeters of polygons can be measured in the same way by working from one polygon node, around the polygon, and back to the same node. Measurement of line lengths is a common task in GIS contexts. As an example, applications concerning road networks (see Chapter 6) often make use of information on the length of road networks.

4.4 Measuring areas

Measurement of areas with raster grids is straightforward. If n cells belong to a given class then the area of a cell (given by the spatial resolution squared) is simply multiplied by n to get the total area covered by pixels in that class. Measurement of the areas of vector polygons is outlined in the following section. Many applications require information on the areas of zones. As an example, to compute population density in an area both the total population and the area of the zone are required.

4.4.1 Areas of polygons

The area of a polygon can be calculated by:

$$A = 0.5 \times \sum_{i=1}^{n} y_i \times (x_{i+1} - x_{i-1}) \qquad (4.2)$$

where x_i and y_i are the x and y locations for node i. In Figure 4.1 a simple polygon feature is shown. The x and y coordinates of its nodes and the calculations following the equation are given in Table 4.1. Note that x_{i+1} refers to the next node in the list and x_{i-1} is the previous one. For the first node (node 1) the previous node is the last node in the list (in this example, node 5).

As an example, we take the y coordinate of node 1 and multiply it by the product of the x coordinate of the next node (obviously, node 2) minus the x coordinate of the previous node (node 5 in this case). Next, we take the y coordinate of node 2 and multiply it by the product of the x coordinate of the next node (node 3) minus the x coordinate of the previous node (node 1 in this case). This is done for each node and the results summed and multiplied by 0.5. Note that the procedure should be followed in a clockwise direction, if it isn't then the area returned will be negative.

In this case the area, A, is given by $0.5 \times 48 = 24$.

The procedure works for any polygon, whatever its degree of complexity. Calculation of areas of polygons is also demonstrated by Kitchin and Tate (2000) and Wise (2002).

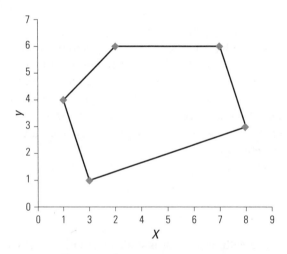

Figure 4.1 Simple polygon feature.

Table 4.1 Simple polygon nodes and area calculations

Node	x_i	y_i	$y_i \times (x_{i+1} - x_{i-1})$
1	2	1	$1 \times (1-8) = -7$
2	1	4	$4 \times (3-2) = 4$
3	3	6	$6 \times (7-1) = 36$
4	7	6	$6 \times (8-3) = 30$
5	8	3	$3 \times (2-7) = -15$
		Sum	48

4.5 Distances from objects: buffers

In applications where selection of objects within a set distance of other objects is the concern, generation of a buffer polygon is a likely step. Buffers are widely used in site-selection projects and in many other contexts where straight line (Euclidean) distances are meaningful. Cases where distance is more logically measured along networks (as would be the case when the distance between places by road is measured), rather than as a straight line between start and end points, are dealt with in Chapter 6.

4.5.1 Vector buffers

A buffer polygon represents the area within a specified distance of an object—that is, if the buffer is computed for a distance of 5 km, the buffer polygon represents a distance of 5 km from the object of interest. Figure 4.2 gives an example of a buffer polygon around a linear feature. The object could also be a point (in which case, the buffer is a circle) or a polygon. Overlay operators, which could be used to identify areas or objects falling within buffer polygons, are discussed in Chapter 5.

One method for generating buffers entails moving a circle with the required radius along the feature to be buffered. The internal boundaries of the overlapping circles can then be dissolved (see Section 4.10 for a discussion about dissolving internal boundaries). Often, buffers are computed for several distance bands and items that fall within each band can be identified using an overlay operator, as detailed in the following chapter. Such an operation is routine in most GIS software. It is also possible to vary the width of the buffer to take into account specific local characteristics. For example, buffers may be wider in areas with steep slopes or in areas that are environmentally sensitive.

4.5.2 Raster proximity

Buffers can also be represented as raster grids. For an input of cells that are coded as locations of interest, a buffer grid can be generated. In such an output, cells within the specified distance of the objects will be given one value (e.g. 1) while cells outside of that area will be given another value (e.g. 0). With the raster model, proximity is often measured directly and each cell in the output records the distance from the cell or cells of interest in the input grid. Figure 4.3A shows such a grid with cell values representing the distance of cells from a linear feature running from the top left to the bottom right of the grid—the feature corresponds to the cells with zero distance

Figure 4.2 Buffer (polygon with light line) around a linear feature (heavy line).

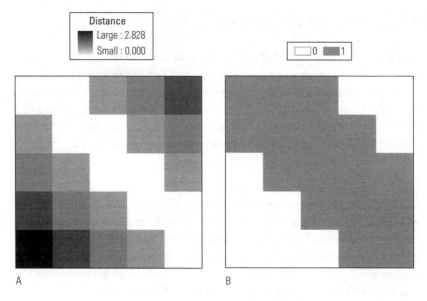

Figure 4.3 Distances from linear feature: (A) distance, (B) binary map for distances less than or equal to 1.4 units (value of 1) or greater (0).

values. A buffer can easily be generated from this grid using a simple classification procedure and Figure 4.3B shows a buffer for distances of less than or equal to 1.4 units. With this approach, all cells with distances of less than, or equal to, the specified amount are coded '1' and all cells with distances greater than this value are coded '0'. It is then straightforward to select all cells in a second image which fall within the buffer defined in the first image. Section 10.2 shows how this can be done. Note that a raster proximity map can be generated directly from vector data, as well as from particular cells in another raster grid.

The following section deals with a key concept in spatial data analysis, the moving window.

4.6 Moving windows: basic statistics in subregions

In many cases, spatial variables have different properties at different locations (for example, values tend to be large in some areas and small in others). In such cases, it is useful to be able to account for these differences and moving windows offer one solution. The idea of the moving window is core to spatial data analysis. In simple terms a moving window represents a region covering part of the entire study area and this region or 'window' is moved from one location to another. Usually, the window is circular or square in shape. In many applications, the window moves in regular steps across the study region, with some operation (e.g. the calculation of the mean average of values in the window) conducted at each location. Such an approach to computing

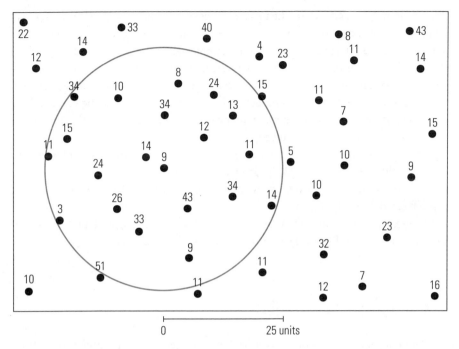

Figure 4.4 Moving window centred on one point: radius of 25 units.

the mean average can be given as follows, for a moving window which is a circle with a radius of 25 units (e.g. metres):

1. Go to location i.

2. Calculate the mean average of all values around location i that fall within 25 m of that location.

3. Make $i=i+1$ (the next location is often at some fixed distance away and in a predetermined direction; the locations may be nodes of a regular grid) and go to step 1.

In Figure 4.4, a window with a radius of 25 units is centred on one observation. The window could be centred anywhere in the study region—the location of an observation or elsewhere. In this case, the window contains 20 values and the sum of these values is 381. The mean of these values is $381/20 = 19.05$. The approach is straightforward to apply whether the data are regularly spaced (like a grid) or irregularly spaced (like, for example, rain gauges tend to be).

Moving windows are used widely in image processing. Usually each cell in an image (raster grid) is visited in turn and some statistic is computed using that cell and its immediate neighbours (note that this is termed a 'focal operator', as described in Sections 10.3 and 10.4). The mean in a moving window might be used to smooth an image (reduce the effect of local outliers) or the standard deviation in a moving window could be used to highlight the edges of features in an image. A focal operator is illustrated in Figure 4.5, where the mean average of neighbouring pixels is computed. Note that when such procedures are employed, the output image often has fewer rows and columns than the input image. In the example below, the window is 3×3 pixels in size and the mean is only computed where there are neighbours on all sides of a pixel. When the window is centred on pixels at the edge of the image, there are fewer than 3×3 pixels and so no value is computed. The moving window statistic could still be calculated from the smaller number of pixels, but in many cases the procedure employed in this example is followed.

In the case of position 1, the value in the centre of the window is 42 and its neighbouring values are 45, 44, 44, 43, 39, 38, 32, and 34. Adding these values together and dividing the sum by nine gives a value (the mean average) of 40.11, as shown in the top-left cell of the output grid.

The next section extends the moving window idea by treating each of the observations in the window differently according to where they are located.

4.7 Geographical weights

The tendency for observations close together in space to be more similar than observations that are separated by larger distances (see Section 4.1) is often accounted for in spatial analyses. For example, a summary statistic computed in a moving window of a particular size may be based equally on all of the data in the window at a certain position. Alternatively, observations close to the centre of the window may be given more weight (or influence). Logically this is sensible: if the summary statistic is allocated to a point in the centre of the window, it is sensible to allow close-by observations to have most influence on the estimated statistic at that location since these close-by values are most likely to be similar. The objective is to obtain a more reliable statistic as distance to neighbours is taken into account.

Weights can be based on adjacency or they can, for example, be a function of distance. The example application of the Moran's I statistic in the following section uses adjacency: neighbouring cells are given a weight of one while all other cells are given a weight of zero, i.e. they are not included in the calculations. Alternatively, all cells (or points/areas) or some subset could be used in the calculations but with larger weights given to cells closer to the cell of interest. There is a large variety of weighting functions that determine how much weight should be given to observations as a function of distance. A simple linear weighting function could be used whereby an observation twice as far away receives half as much weight, e.g. an observation at 10 km receives twice as much weight as an observation 20 km away. In practice, more sophisticated schemes are used for most applications. One well-known weighting function is based on taking the inverse of the squared distance from the location of interest (following the inverse square law). In other words, the weight is a function of (is dependent on) the inverse squared distance, d^{-2} (this can be obtained with $1/d^2$, as detailed below, and see Section 9.5 for an application of this weighting function). Whatever distance decay weighting

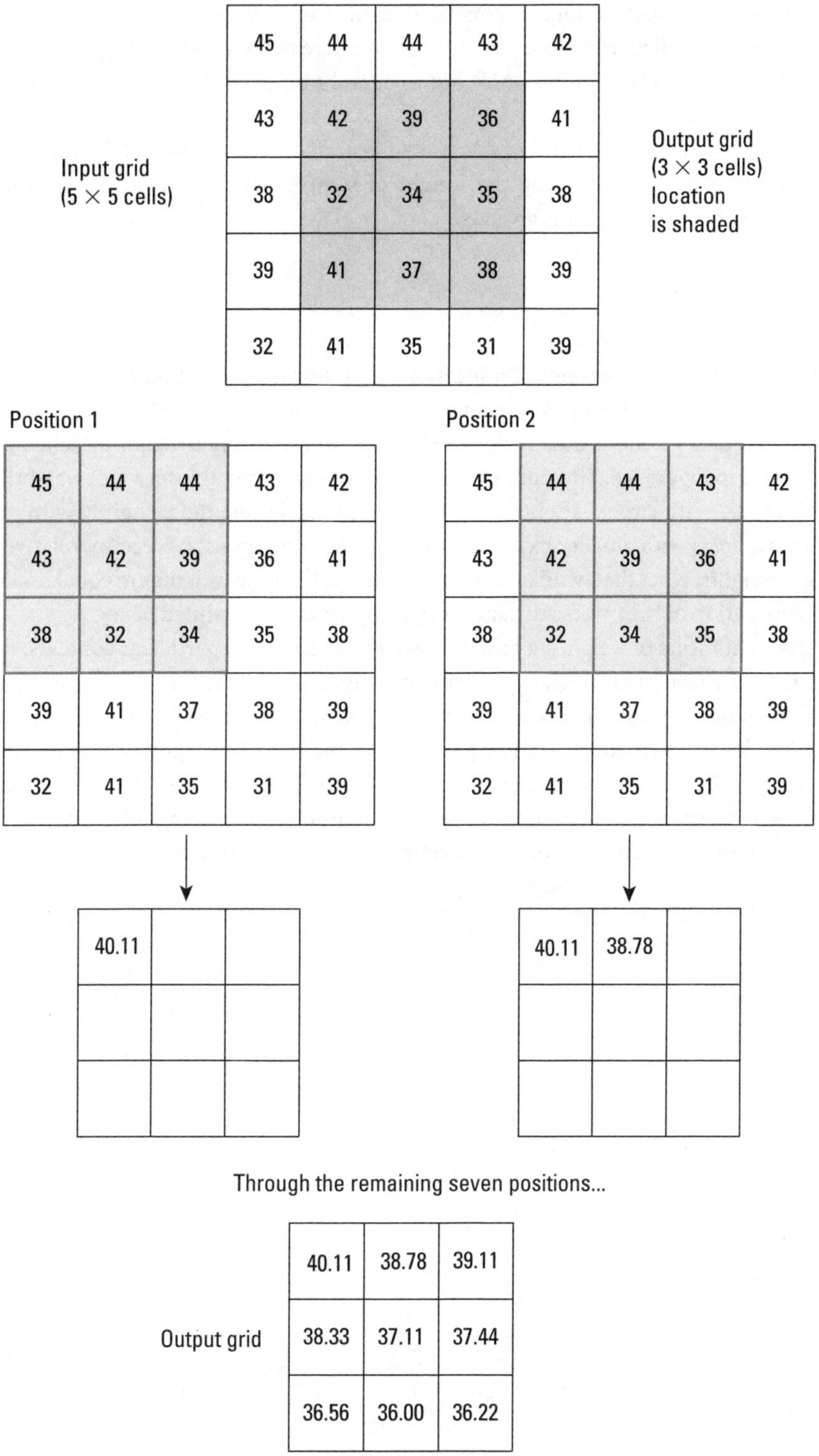

Figure 4.5 Mean average computed for a 3×3 pixel moving window.

scheme is used, observations at smaller distances have larger weights than observations at larger distances from the location of interest. Some other weighting functions are described later in this book, but a general summary of distance weighting is provided below.

The inverse distance weighting scheme is illustrated now. The weight for location i can be given by w_{ij}, indicating the weight of sample i with respect to location j. The inverse distance weight is given by:

$$w_{ij} = d_{ij}^{-k} \tag{4.3}$$

which indicates that the weight for location i with respect to location j is obtained by raising the distance d between locations i and j (i.e. d_{ij}) to the power $-k$. As noted above in the case of $k=2$, this is obtained with $1/d^k$. The inverse distance weighting scheme is illustrated in Figure 4.6. The value of the exponent determines the degree of weighting by distance. With larger exponent values, the weights decline more sharply with distance, whereas with smaller exponent values distant observations receive, relatively, larger weights. Note that, with an exponent of zero, all of the weights are equal to one. An application of the inverse distance weighting scheme is outlined below.

Different forms of weighting scheme have found favour in particular contexts. For example, the Gaussian weighting scheme described in Section 8.4 has been used for weighting observations as a part of a method called geographically weighted regression, which is detailed in Section 8.5.3, while the quartic weighting scheme (see Section 7.3.2) has been used for point pattern analysis. Inverse distance weighting is the basis of a spatial interpolation method (a method for predicting values at unsampled locations), which is discussed in Section 9.5 and is illustrated briefly here.

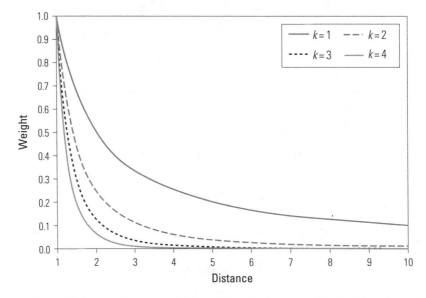

Figure 4.6 Inverse distance weighting scheme for exponents (k) of 1, 2, 3, and 4.

Any standard statistic can be geographically weighted (see Fotheringham *et al.* (2002) for more information). As an example of a geographical weighting scheme in practice, obtaining the locally weighted mean using inverse distance is illustrated below. The locally weighted mean is given by:

$$\overline{z}_i = \frac{\sum_{j=1}^{n} z_j w_{ij}}{\sum_{j=1}^{n} w_{ij}} \tag{4.4}$$

Recall from Equation 3.3 that $\overline{z}$ indicates the mean of z, so $\overline{z}_i$ indicates the mean at location i; w_{ij} indicates, as before, the weight for the distance between observations i and j. In the case where all the weights are one, Equation 4.4 corresponds to the standard mean average (it is then the sum of the values divided by the number of observations).

In Table 4.2, a set of observations (which can be treated as measurements of precipitation in millimetres for illustrative purposes) and their distance from a fixed location are given. In this case, the value at the fixed location is unknown and it will be

Table 4.2 Observations (j), distance from observation 1 (d_{ij}), weights (w_{ij}), and weights multiplied by values ($z_j w_{ij}$)

j	d_{ij}	z_j	$k=1$ w_{ij}	$z_j w_{ij}$	$k=2$ w_{ij}	$z_j w_{ij}$	$k=3$ w_{ij}	$z_j w_{ij}$
1	4.404	14	0.227	3.179	0.052	0.722	0.012	0.164
2	9.699	43	0.103	4.434	0.011	0.457	0.001	0.047
3	10.408	12	0.096	1.153	0.009	0.111	0.001	0.011
4	10.871	34	0.092	3.127	0.008	0.288	0.001	0.026
5	12.958	26	0.077	2.007	0.006	0.155	0.000	0.012
6	13.959	24	0.072	1.719	0.005	0.123	0.000	0.009
7	14.066	33	0.071	2.346	0.005	0.167	0.000	0.012
8	15.506	34	0.064	2.193	0.004	0.141	0.000	0.009
9	17.256	10	0.058	0.579	0.003	0.034	0.000	0.002
10	17.606	8	0.057	0.454	0.003	0.026	0.000	0.001
11	18.018	13	0.055	0.721	0.003	0.040	0.000	0.002
12	18.025	11	0.055	0.610	0.003	0.034	0.000	0.002
13	18.285	24	0.055	1.313	0.003	0.072	0.000	0.004
14	19.253	9	0.052	0.467	0.003	0.024	0.000	0.001
15	21.335	15	0.047	0.703	0.002	0.033	0.000	0.002
16	23.845	14	0.042	0.587	0.002	0.025	0.000	0.001
17	23.988	34	0.042	1.417	0.002	0.059	0.000	0.002
18	24.464	3	0.041	0.123	0.002	0.005	0.000	0.000
19	24.522	11	0.041	0.449	0.002	0.018	0.000	0.001
	Sum	372	1.347	27.582	0.128	2.533	0.017	0.309
	Mean	19.579		20.474		19.844		17.804

Weights are obtained using the inverse distance weighting scheme with exponents of 1, 2, and 3.

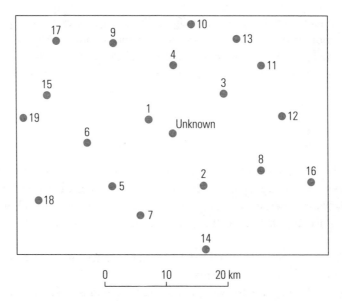

Figure 4.7 Locations of observations listed in Table 4.2.

predicted using inverse distance weighting. Figure 4.7 shows the locations of the observations, along with the location for which a prediction will be made. The weights, obtained using the inverse distance weighting scheme detailed above (with exponents of 1, 2, and 3), are given for each distance. The weights for each location are then multiplied by the value at that location. As an example, following Equation 4.4 (and with weights as defined in Equation 4.3) for an exponent of 1, the products of the multiplications are summed, giving a value of 27.582. The weight values are also summed, giving a value of 1.347. The weighted mean is then obtained as $^{27.582}/_{1.347} = 20.474$. The mean obtained without geographical weighting (i.e. with all weights equal to 1) is 19.579.

Weighted means (or other statistics) can be calculated anywhere: at the location of an observation or anywhere else. Section 8.4 demonstrates the geographically weighted mean using another weighting scheme. Fotheringham *et al.* (2002) discuss a range of geographically weighted statistics. Geographical weights will be encountered throughout this book.

Selection of a weighting scheme is usually arbitrary, but there may be characteristics of a data set that guide selection. In essence, the choice of weighting function (and, where relevant, parameters like the bandwidth, see Sections 7.3 and 8.4) should be determined either through experimentation (e.g. for interpolation, which weighting function leads to the most accurate predictions) or knowledge of the process of interest—if we know something about the scale of variation (see Section 2.7) this may inform our choice of weighting scheme.

The focus now moves from measurement of distances to characterizing the spatial structure of values (i.e. how similar neighbouring values are to one another).

4.8 Spatial dependence and spatial autocorrelation

A key concern in spatial data analysis is to examine spatial patterning in the variable or variables of interest. For example, are values of a particular variable large in some areas and small in others? Also, do similar values tend to cluster or are values visually erratic? The term 'spatial dependence' refers to the dependence of neighbouring values on one another (Haining, 2003). As outlined at the start of this chapter, the basis of spatial dependence is that values close together in space tend to be more similar than those that are farther apart. The 'first law of geography' (Tobler, 1970) is a key concept in geography in general and spatial data analysis in particular. In the context of statistical measurement, this idea is related to spatial autocorrelation—the degree to which a variable is spatially correlated with itself. A measure of spatial autocorrelation may suggest spatial dependence (i.e. neighbouring values are similar—positive spatial autocorrelation) or spatial independence (neighbouring values are dissimilar—negative spatial autocorrelation).

There is a range of measures of spatial autocorrelation. The joins count approach is one means of summarizing the tendency of neighbouring observations to be the same (O'Sullivan and Unwin, 2002). The measure of spatial autocorrelation encountered most frequently in the spatial analysis literature is the I coefficient proposed by Moran (Moran, 1950; Cliff and Ord, 1973). It is given by

$$I = \frac{n\sum_{i=1}^{n}\sum_{j=1}^{n}w_{ij}(y_i - \overline{y})(y_j - \overline{y})}{\left(\sum_{i=1}^{n}(y_i - \overline{y})^2\right)\left(\sum_{i=1}^{n}\sum_{j=1}^{n}w_{ij}\right)} \qquad (4.5)$$

where the values y_i (of which there are n) have the mean $\overline{y}$ and the proximity between locations i and j is given by w_{ij}. As before, this is a geographical weight and is often set to 1 when locations i and j are neighbours and 0 when they are not. Note that here y is a data value and not a coordinate. Elsewhere in this book, z is used to represent data values but y is used here to distinguish the use of z as a deviation of y from its mean in the local spatial autocorrelation measures detailed in Section 8.4.1. Equation 4.5 includes double summations (note that single summation was introduced with respect to Equation 3.1), that is:

$$\sum_{i=1}^{n}\sum_{j=1}^{n}$$

This means start with $i=1$ and $j=1$, next, $i=1$ and $j=2$ then $i=1$ and $j=3$, and so on until $j=n$. After that point, $i=2$ and we work through all values of j until all combinations of i and j have been accounted for. At each stage, the computed values are added to the values obtained previously. In this way all combinations of i and j are included. With the numerator of Equation 4.5

$$n\sum_{i=1}^{n}\sum_{j=1}^{n}w_{ij}(y_i - \overline{y})(y_j - \overline{y})$$

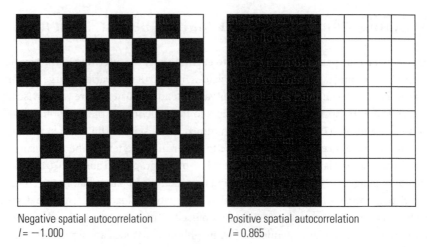

Negative spatial autocorrelation
$I = -1.000$

Positive spatial autocorrelation
$I = 0.865$

Figure 4.8 Spatial autocorrelation: rook's case contiguity. Black cells have a value of 1, white cells have a value of 0.

the calculations are computed for every combination of i and j and the results of each calculation are added together, with the product being multiplied by the weight w_{ij}. The procedure to calculate the Moran's I value is demonstrated below. Negative values of I indicate negative spatial autocorrelation—neighbouring values tend to be different. Positive values of I indicate positive spatial autocorrelation—neighbouring values tend to be similar. Values of I close to zero indicate that there is no structure. This section first outlines the basic concepts, then gives a small, fully worked numerical example, and finally gives an example using a larger grid.

Values of I for two different grids are given in Figure 4.8. These examples were computed using the package GeoDa (Anselin *et al.*, 2006) and it should be noted that the results are slightly different to those that would be calculated using Equation 4.5 as GeoDa modifies the form of the weights (in that package the weights are what are termed 'row standardized', i.e. they sum to 1), but that isn't a concern here. In this example, black cells have a value of 1 while white cells have a value of 0. For this example, cells which share an edge with another cell are compared and not cells which share only corners. Through the analogy with movement of pieces in chess, this is called rook's case contiguity. Where cells which share corners (i.e. cells connected diagonally) are also included, this is called queen's case contiguity. Rook's case and queen's case contiguity are illustrated in Figure 4.9. In the case of irregularly shaped zones, rook's case contiguity and queen's case contiguity can also be used, with the latter including zones that are connected only by vertices as well as by edges, while the former includes only zones joined by edges. In packages such as GeoDa, different weighting functions (e.g. rook's case and queen's case) can be used and it is necessary to consider which is used and how this may impact on the results. As well as rook and queen contiguity, other weighting schemes can be used (see Section 4.7).

The case on the left of Figure 4.8 indicates negative spatial autocorrelation—all neighbours of a given cell are, using rook's case contiguity, different to that cell. In the

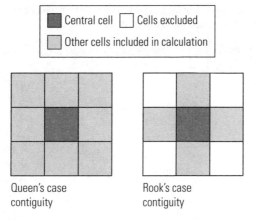

Figure 4.9 Queen's case contiguity and rook's case contiguity.

case on the right, the neighbours of most cells have the same value and, therefore, the values are positively spatially autocorrelated. The only exceptions are the cells in the middle two columns of the grid, which have different values.

Using a small grid of values, Moran's *I* is illustrated below. Note that the method is equally applicable to zones with irregular forms. The sample grid is:

```
7    8   11
11   9   10
11   12   9
```

Values that are next to one another along rows or columns (e.g. 7 and 8 or 9 and 10) will be counted as neighbours as will those that are next to one another diagonally (e.g. 8 and 10). As noted above, this is called queen's case contiguity.

First we will calculate $(y_i - \bar{y})(y_j - \bar{y})$—that is, the difference of each value from the mean multiplied by the difference between each neighbouring value and the mean. For example, the value 7 (top left cell) minus the mean (9.778) is −2.778. One of the neighbours of this cell has the value 8 and its difference from the mean is −1.778. We then multiply the two differences together, giving 4.938 (see the last entry of the top row in Table 4.3). This is done for every cell and its neighbours, as shown in Table 4.3. Note that Table 4.3 includes only cells that are neighbours (so the weight in each case is 1). The sum of the products, $\sum_{i=1}^{n}\sum_{j=1}^{n}w_{ij}(y_i - \bar{y})(y_j - \bar{y})$, is 3.975.

Next we will calculate $(y_i - \bar{y})^2$, the squared difference between each value and the mean. The results are shown in Table 4.4. The sum of squared differences from the mean $(\sum_{i=1}^{n}(y_i - \bar{y})^2)$ is 21.556.

There are nine observations, $\sum_{i=1}^{n}\sum_{j=1}^{n}w_{ij}(y_i - \bar{y})(y_j - \bar{y}) = 3.975$, the sum of squared differences from the mean is 21.556 and there are 20 adjacencies (the number of rows in Table 4.3 is twice the number of adjacencies). As an example, the cells with values 7 and 8 are neighbours (i.e. they are adjacent to one another). Each adjacency (like all the others) is counted twice as we have 7 paired with 8 and 8 paired with 7.

Table 4.3 Value, value − mean (9.778), neighbour value, neighbour value − mean, product (two differences from mean multiplied together)

Value y_i	Value − mean $y_i - \bar{y}$	Neighbour value y_j	Neighbour − mean $y_j - \bar{y}$	Product $(y_i - \bar{y})(y_j - \bar{y})$
7	−2.778	8	−1.778	4.938
7	−2.778	9	−0.778	2.160
7	−2.778	11	1.222	−3.395
11	1.222	7	−2.778	−3.395
11	1.222	8	−1.778	−2.173
11	1.222	9	−0.778	−0.951
11	1.222	12	2.222	2.716
11	1.222	11	1.222	1.494
11	1.222	11	1.222	1.494
11	1.222	9	−0.778	−0.951
11	1.222	12	2.222	2.716
8	−1.778	7	−2.778	4.938
8	−1.778	11	1.222	−2.173
8	−1.778	9	−0.778	1.383
8	−1.778	10	0.222	−0.395
8	−1.778	11	1.222	−2.173
9	−0.778	7	−2.778	2.160
9	−0.778	8	−1.778	1.383
9	−0.778	11	1.222	−0.951
9	−0.778	10	0.222	−0.173
9	−0.778	9	−0.778	0.605
9	−0.778	12	2.222	−1.728
9	−0.778	11	1.222	−0.951
9	−0.778	11	1.222	−0.951
12	2.222	11	1.222	2.716
12	2.222	11	1.222	2.716
12	2.222	9	−0.778	−1.728
12	2.222	10	0.222	0.494
12	2.222	9	−0.778	−1.728
11	1.222	8	−1.778	−2.173
11	1.222	9	−0.778	−0.951
11	1.222	10	0.222	0.272
10	0.222	11	1.222	0.272
10	0.222	8	−1.778	−0.395
10	0.222	9	−0.778	−0.173
10	0.222	12	2.222	0.494
10	0.222	9	−0.778	−0.173
9	−0.778	12	2.222	−1.728
9	−0.778	9	−0.778	0.605
9	−0.778	10	0.222	−0.173

Table 4.4 Values, difference from the mean (9.778) and the squared differences

Value y_i	Difference $y_i - \bar{y}$	Squared difference $(y_i - \bar{y})^2$
7	−2.778	7.716
11	1.222	1.494
11	1.222	1.494
8	−1.778	3.160
9	−0.778	0.605
12	2.222	4.938
11	1.222	1.494
10	0.222	0.049
9	−0.778	0.605

The sum of the weights therefore, $\sum_{i=1}^{n}\sum_{j=1}^{n}w_{ij}$ (the right-hand side of the denominator of Equation 4.5), is 40. Moran's I is computed by:

$$I = \frac{9 \times 3.975}{21.556 \times 40} = \frac{35.778}{862.222} = 0.041$$

Since this value is close to 0, this indicates that neighbouring values in the example do not tend to be similar. Another example is given which indicates negative spatial autocorrelation. For the grid:

```
 7   8  6
10  14  7
 6  11  9
```

This leads to:

$$I = \frac{9 \times -66.889}{56.000 \times 40} = \frac{-602.000}{2240.000} = -0.269$$

In this case, neighbouring values tend to be dissimilar, thus no clustering of like values is suggested.

It will be obvious that increasing the size of the grid will necessitate the use of a computer to obtain a value of Moran's I. A further example of I is given in Figure 4.10. Like the previous example (Figure 4.9), the calculations were conducted using GeoDa. Figure 4.10 shows the potential difference in results obtained using rook's case (using only cells joined by edges) or queen's case (using cells joined by edges or by corners/vertices).

As noted in Section 3.5, spatial autocorrelation has an impact on standard statistical procedures. In essence, a large sample size gives greater confidence in the inferences we make than a small sample size; this is intuitively obvious. If neighbouring values are similar then the observations are considered dependent on one another. A practical implication is that the degree of confidence in our results will be smaller than the

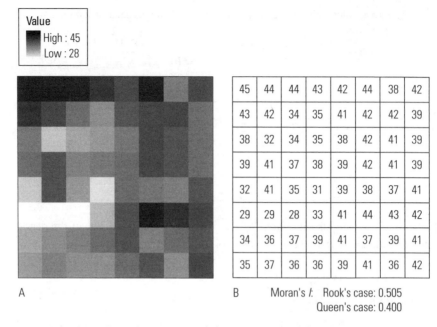

45	44	44	43	42	44	38	42
43	42	34	35	41	42	42	39
38	32	34	35	38	42	41	39
39	41	37	38	39	42	41	39
32	41	35	31	39	38	37	41
29	29	28	33	41	44	43	42
34	36	37	39	41	37	39	41
35	37	36	36	39	41	36	42

A

B   Moran's *I*: Rook's case: 0.505
                      Queen's case: 0.400

Figure 4.10 Example raster using (A) grey scales and (B) numerical values: spatial autocorrelation for queen's case and rook's case contiguity.

sample size suggests, since the observations are not independent of one another. As noted previously, it is important to remember that we must be wary of such problems when applying standard statistical procedures (e.g. significance tests) in the analysis of spatial data (see Rogerson (2006) for a discussion about this topic).

There is a variety of other methods for measuring the degree of spatial autocorrelation. The purpose of this chapter is only to introduce topics and the analysis of spatial autocorrelation is explored further in Chapter 8.

4.9 The ecological fallacy and the modifiable areal unit problem

It is often necessary to work with spatially aggregated data, for example census zones or cells in remotely sensed images. Such zones are unlikely to be internally homogeneous. For example, a cell in a remotely sensed image has only one value, but in the real world there may be several features in the area covered by the cell. In words, the variation within the cell (or other area) is lost if the area is larger than the individual features it contains. This section explores two sets of concepts that relate to such issues. These are the ecological fallacy and the modifiable areal unit problem.

The ecological fallacy refers to the problem of making inferences about individuals from aggregate data. For example, not all people in one census zone are likely to share the same characteristics. The majority of people in a census zone may be wealthy,

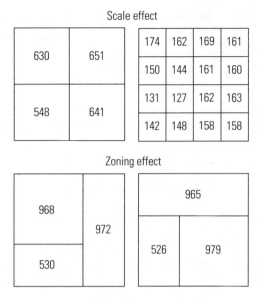

Figure 4.11 The scale and zoning effects.

but if there is a housing estate just inside one edge of the zone then clearly generalizations about the population of the zone may be unsound. Geographical space can be divided in an infinite number of ways. In practice, it is often the case that data aggregated over only one set of areal units are available. It may be that no set of zones has intrinsic meaning about the underlying populations and that the units are 'modifiable'. This problem is sometimes termed the 'modifiable areal unit problem' (MAUP) (Openshaw and Taylor, 1979). The MAUP is composed of two parts:

The scale effect   Statistical analyses based on data aggregated over areas of different sizes will produce different results.

The zoning effect   Two sets of zones can have the same or similar areas but very different forms and analyses based on two such sets of zones may vary.

The scale effect and the zoning effect are illustrated in Figure 4.11. In this example, the values (representing numbers of people in each area) shown in Figure 4.10 have been aggregated into new larger zones.

Note that different terms are used in the literature to refer to the two component parts of the MAUP. For example, for what is termed here the 'scale effect' the term 'aggregation effect' is sometimes used (e.g. Atkinson and Tate, 2001; Waller and Gotway, 2004). To add to the confusion, the zoning effect is sometimes also referred to as the aggregation effect (e.g. Openshaw and Taylor, 1979). Other terms are also encountered for both components of the MAUP but, irrespective of this, the key distinction between the two components is that one deals with the size of zones (here called the scale effect) and the other with changes in their shape or position when the size of zones is the same or similar (here called the zoning effect).

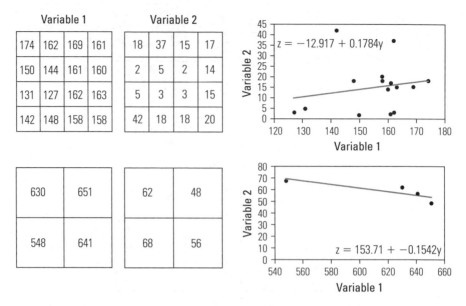

Figure 4.12 Two scatter plots and fitted lines for different aggregations of the same values.

Wong (1997) explores changes in measures of residential segregation (i.e. the degree to which members of different groups live in different areas). Wong argues that, if the counts of the population group are negatively spatially autocorrelated (neighbouring zones are dissimilar), using zones of different sizes will result in different segregation measure values. Conversely, if the counts are positively spatially autocorrelated (neighbouring zones are similar), then using zones of different sizes will make little difference for zones smaller than the area (or scale) over which counts are positively autocorrelated. In general, the degree of spatial autocorrelation is important when considering the effect of changing the zonal system used.

Figure 4.12 gives an artificial example of the potential effects of altering the aggregation of values. In this case, two sets of variables are given for two different aggregations. Regression of one variable on the other is then conducted using the two sets of aggregations. Note that the sample size is *very* small and this example is used purely for illustrative purposes. Recall from Section 3.4 that assessing sample size is important when considering regression results. In the example, the slope changes from positive to negative as the values are aggregated over larger units. Even where the sign doesn't change, the effects of changing the size or form of zones may be highly significant. Openshaw and Taylor (1979) explore the issue using regression for many different zonal systems and they demonstrate large differences where the form of zones varies but their number is the same (the zoning effect) and where the number of zones varies (the scale effect). Note that r^2 tends to increase with increased aggregation, and this would clearly be the case given the example in Figure 4.12.

In summary, the potential impact of the size and shape of zones on results should be considered and it should be remembered that any pattern apparent in mapped areal data may be due as much to the zoning system used as to the underlying distribution

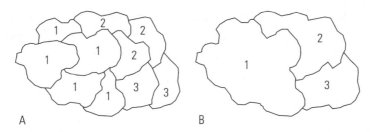

Figure 4.13 (A) Original polygons and (B) polygons with internal boundaries dissolved.

of the variable (Martin, 1996). Raster cells can also be conceptualized as zones and the spatial resolution of a raster will similarly determine results.

4.10 Merging polygons

Related to the previous theme, a common operation in GIS contexts is the merging of polygons with common attributes (this is sometimes termed the 'dissolve operator'). For example, if we have sets of areas all of which have the same area code then we may wish to merge those areas where their boundaries are adjacent. If there are several zones within one larger administrative area, all of which have the same zone identity, then if we dissolve the internal boundaries (i.e. the boundaries of the small zones) we are left with the boundaries of the larger administrative areas. Figure 4.13 gives an example.

Summary

This chapter was concerned with some very basic operations, but such methods form the core of many GIS-based analyses. Measurement of straight line (Euclidean) distances, an initial focus of this chapter, is sensible in many contexts; in others they may not be meaningful. For example, the movement of airborne pollutants is a function of various factors, such as wind direction, and simple distance may explain such a process only to a very limited degree. Also, buffers may only be useful in particular contexts. If we are concerned with accessibility of a particular location in a populated area then measuring the straight line distance of places to that location may not be very helpful. Instead, we may wish to measure distances (and perhaps calculate travel times) along a road network (perhaps using the approach detailed in Section 6.5). Other ways of accounting for the 'cost' of moving from one place to another include cost (or friction) surfaces and this idea is discussed in Section 10.6. Following the discussion about distances, areas, and buffers, some particular concepts in spatial analysis were introduced. These included moving windows, geographical weights, spatial dependence and spatial autocorrelation, the ecological fallacy, and the MAUP. At least some of these concepts are central to all analyses of spatial data and these ideas will be revisited throughout the remainder of this book.

Further reading

Most standard introductions to GIS provide accounts of measurement of distances and areas. General summaries are provided by, for example, Burrough and McDonnell (1998) and Heywood *et al.* (2006). Chou (1997) and O'Sullivan and Unwin (2002) provide more in-depth accounts of the key ideas. Wise (2002) outlines an algorithm for measurement of polygon areas. Key spatial data analysis concepts such as moving windows and geographical weighting are discussed by O'Sullivan and Unwin (2002) and Lloyd (2006). The MAUP is outlined in some detail by Openshaw (1984). Chapters in the book edited by Tate and Atkinson (2001) introduce some key concepts and present case studies relating to the issue of spatial scale in GIScience. By carefully working through this chapter, and the two which preceded it, readers should have developed the essential background necessary to make use of the rest of this book.

➡ The next chapter is concerned with analysis of discrete objects. Specifically, it deals with overlay operators, which are used to identify overlaps between spatial objects.

5

Combining data layers

5.1 Introduction

This chapter is concerned with the nature of connections between objects in terms of overlap or proximity between different objects. Operators that deal with overlaps between objects require information on where specific features overlap in space—the overlap of lines is termed the 'intersection'. Such operators provide the basis of multi-criteria decision analysis (MCDA, see Section 5.3) whereby information in multiple data layers is taken into account. Overlay operators are a key component of GIS software packages and they enable the identification of features that share the same geographical space (whether points, lines, or areas). The main focus in this chapter is on vector data; Section 10.2 deals briefly with raster overlay.

Exploring connections between objects may provide the basis of many different kinds of analyses. At a simple level, for example, there may be a need to identify which areas border another area (i.e. which areas are contiguous). Using topological structures, as detailed in Section 2.2.3, this kind of information may be obtained directly. Often the concern will be to identify which objects contain other objects or are contained by other objects (e.g. which local government boundary contains which houses). In many cases, connectivity between different kinds of objects is of interest. Finding the common areas from two sets of polygon features (e.g. highly populated areas and highly polluted areas) is one kind of application that is frequently encountered in GIS contexts. This chapter is concerned with overlaps between features in different data layers.

5.2 Multiple features: overlays

Overlay operators combine information in various ways from two or more sets of spatial data. Such operators are concerned with inclusion and with overlap or intersection (Burrough and McDonnell, 1998). For properties (e.g. areas) A and B, the two can be defined as follows:

Inclusion   Is A contained within B? Operators transfer attributes to the features contained or that contain other features. For example, if point A is contained within polygon B, then the point is labelled as belonging to area B and attribute information from B is transferred to A.

Intersection   Do A and B overlap? This leads to the creation of new spatial features. For example, if two areas overlap then at the points where the edges overlap new nodes are created and overlapping areas in the output contain the attributes of both input layers (in this example, they have type A AND type B). Inclusion and intersection are both illustrated below.

Overlay operators can be used for polygon overlay, line-in-polygon overlay, point-in-polygon overlay, and to identify overlapping lines. Polygon overlay is a spatial operation that overlays one polygon layer on another to create a new polygon layer. The spatial features of each set of polygons (or a subset) and their polygon attributes are joined in the output layer. Joining polygons enables the use of operations requiring new polygon combinations (e.g. all areas that are both highly populated and highly polluted). Line-in-polygon operations allow the line features to take the attributes of the polygon in which they lie (e.g. identification of which census areas a road passes through). Point-in-polygon overlay transfers the attributes of the polygon in which the point lies to the point (e.g. labelling a house with the administrative region it is contained by). These operations (including point-in-polygon overlay) necessitate identification of line intersections. A means of identifying line intersections is outlined in Appendix D.

5.2.1 Point in polygon

Ascertaining which polygon a point falls in is a frequent problem in GIS contexts and questions such as 'Which disease events are in which administrative areas?' are often posed. There are various ways in which this can be resolved. Laurini and Thompson (1992) outline one approach. The essence of this algorithm is that a line (the 'half line') is drawn from the point to the edge of the map and the number of intersections with lines that are part of the current polygon is counted. The number of intersections will only be odd if the point falls within the boundary of the current polygon. This procedure is illustrated in Figure 5.1. In this example the half line crosses the boundary of the polygon in which it sits three times, confirming that is where it is located. The search process can be accelerated through the use of a minimum enclosing rectangle (MER, as defined in Appendix D) to ascertain if the point can possibly be contained within a given polygon (i.e. if the point is within the MER then it may also be within the polygon contained by the MER). Once polygons are excluded in this way, the remaining candidates can be tested using the line segments (see Appendix D for a testing procedure) until the relevant polygon is identified. Heywood *et al.* (2006) detail problem cases, for example where the point lies on a polygon boundary or the half line

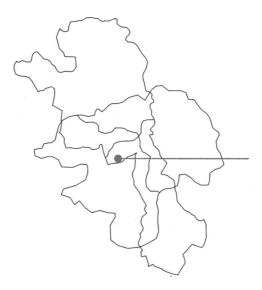

Figure 5.1 Point-in-polygon test example.

coincides with a boundary. Such cases are dealt with through additional stages added to the algorithm employed.

The point-in-polygon operation can be extended to the idea of assessing if an island polygon is contained within another polygon. Firstly a MER can be used to assess if containment of the island polygon by another polygon is possible. Next, a series of horizontal bands is drawn across the polygon that may contain the island polygon. The first and last locations on these lines that intersect the island polygon are then treated as for a point (as shown above). If the number of intersections for each line is odd then the polygon is enclosed completely (Burrough and McDonnell, 1998).

5.2.2 Overlay operators

This section deals with methods that enable answering of questions such as 'How many sites of special scientific interest (SSSI) occur in areas with highly polluted soil?'. This kind of question is well depicted using Boolean logic (see Section 2.11.1). If SSSIs are termed 'A' and highly polluted soil is termed 'B' then the statement A AND B (also given by $A \cap B$) would correspond to our question: all areas fulfilling both criteria are selected. In Appendix D, an approach to the line intersection problem is outlined and this approach can be employed to implement the methods detailed in this section.

Polygon overlay operations are defined and illustrated below. These approaches assume that the features of interest (e.g. polygons representing SSSIs and those representing areas with highly polluted soil) are stored in separate layers. In turn the Union, Intersect, and Identity operations are outlined. These operations are similar, differing only in the spatial features that remain in the output data layer. In all cases, an input layer and a second overlay layer are shown along with the overlay output.

Union overlay joins two sets of polygons and retains all areas from both layers (and the attributes in both layers are kept), so it makes no practical difference which is the

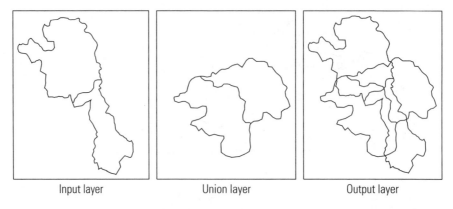

Input layer Union layer Output layer

Figure 5.2 Union operator.

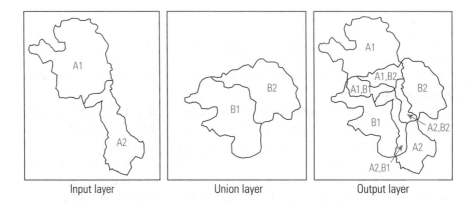

Input layer Union layer Output layer

Figure 5.3 Union operator: transfer of attributes.

input layer and which is the union layer. The operation corresponds to the Boolean OR and is illustrated in Figure 5.2.

The attributes from the two layers are joined as shown in Figure 5.3. In that case, the first layer has polygons that each represent one attribute labelled A1 or A2. Similarly, the polygons in the second layer represent attributes labelled B1 and B2. In the output (union) layer those overlapping areas take both sets of attributes. In this way, new polygons with particular combinations of attributes are generated. Of course, each input polygon may contain many attributes, which will be transferred in the same way.

Intersect overlays points, lines, or polygons on polygons but retains only those portions of the input layer falling within the overlay (intersect) layer features. The intersect overlay corresponds to Boolean AND (for layers A and B, given by $A \cap B$) and is illustrated in Figure 5.4.

For lines, line segments in the input layer that fall within the intersect layer are retained. For points, those points in the input layer that are located within the intersect

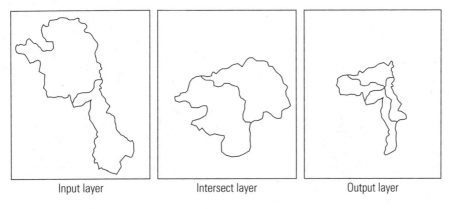

Input layer   Intersect layer   Output layer

Figure 5.4 Intersect operator.

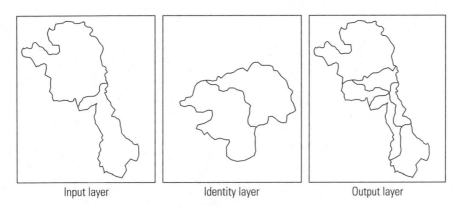

Input layer   Identity layer   Output layer

Figure 5.5 Identity operator.

layer features are retained. Attributes from both layers are transferred to the output layer.

Identity overlays points, lines, or polygons on polygons and keeps all input layer features (with transfer of attributes for both layers). The identity overlay is illustrated in Figure 5.5.

For lines, all input lines are retained but the lines are split and nodes added where the lines overlap with the identity layer features. For points, all input points are retained but the ID numbers of the polygon/s within which points fall are assigned to those points.

5.2.3 'Cookie cutter' operations: erase and clip

'Cookie cutter' operations are widely used—these enable cutting features within a given feature (termed the 'erase operator') or outside a given feature (sometimes termed the 'clip operator'). Unlike the operators described above, attributes are not transferred from both input layers to the output layer—instead only the attributes of the input layer appear in the output layer.

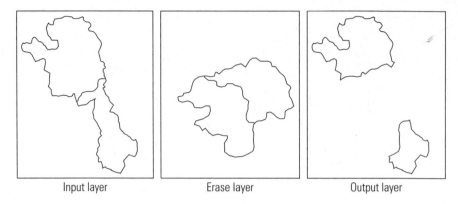

| Input layer | Erase layer | Output layer |

Figure 5.6 Polygon erase.

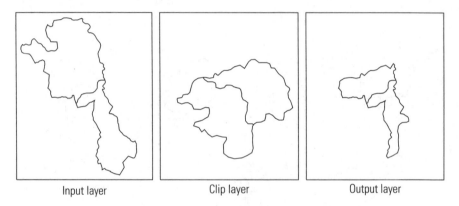

| Input layer | Clip layer | Output layer |

Figure 5.7 Polygon clip.

Erase creates a new layer by overlaying two sets of features. The polygons of the erase layer define the erasing region. Input layer features that are within the erasing region are removed—this is illustrated in Figure 5.6. The erase operator can be used with polygons, lines, or points as inputs.

The clip operator is similar to erase except that the features that are within the clip region are preserved. Like the erase operator, the clip operator can be used with polygons, lines, or points as inputs. The clip operator is illustrated in Figure 5.7.

5.2.4 Applications and problems

Overlays of various kinds are central to many GIS-based projects. A large proportion of applications that utilize multiple spatial attributes make use of overlay procedures. As an example, Sprague *et al.* (2007) are concerned with assessment of the persistence of rice paddies within the Kanto Plain of Japan. In that study, historic (i.e. late nineteenth century) maps were georeferenced to a modern map using features that appear on both the historic maps and modern maps. Overlay procedures were then used to combine rice paddies represented on the historic and the modern maps. It was then possible to identify the common areas of rice paddies on the two maps and to ascertain which areas no longer contain rice paddies.

Combining different data layers may be problematic where polygon boundaries in the input layers are supposed to be identical but in fact differ. The end result is termed 'spurious (or slither) polygons'—(usually) small polygons that represent the difference between the two sets of boundaries. Various approaches exist for removing such spurious polygons. If there is greater confidence in the accuracy of one set of boundaries than the other then the boundaries with greater accuracy may be retained in preference to the other less accurate boundaries.

5.3 Multicriteria decision analysis

GIS allow the integration of diverse data sources and facilitate the exploration of the relationships between them. At a simple level, it may be desired to identify a set of areas that fulfil several dozen criteria (e.g. areas more than a given distance from some feature, with gradual slopes, a particular soil type, and so on). Earlier sections introduced tools, such as buffers (Section 4.5) and overlay operators (previous section), which can be used to help address such issues. This section expands on some key topics that relate to considering multiple criteria in this way. The term 'GIS-based multicriteria decision analysis' (GIS-MCDA) encapsulates the processes involved in using GIS to help decision making using multiple data sources. Malczewski (2006) provides a simple definition of GIS-MCDA as 'a process that transforms and combines geographical data and value judgements (the decision-maker's preferences) to obtain information for decision making' (p. 703). With GIS-MCDA, the relative importance of different criteria can be taken into account. For example, if, in some planning process, accessibility by road is more important than the slope of the terrain then proximity to roads may be given a larger weight than slope as a criterion in the decision-making process.

A means of selecting particular alternatives from a set of available options is a decision rule (Malczewski, 2006). The weighted summation decision rule approach and similar approaches are the most commonly applied in the literature (Malczewski, 2006). Malczewski (1999) outlines one simple additive weighting method. Using this approach each criterion layer is standardized (e.g. by dividing each value in a layer by its maximum value). This is necessary to enable comparison of like with like as direct comparison of, say, distance from roads with slope of terrain would be meaningless—standardization means that all units will be comparable (using the approach suggested above they will all range from 0 to 1; see Heywood *et al.* (2006), pp. 239–240, for another example). Next, weights are determined (this may be quite a subjective procedure). If a layer is to be assigned 40% of the weight this can be expressed as 0.4, with the weights for the other layers totalling 0.6. Then, each of the standardized map layers is multiplied by the weights and the weighted standardized maps are added together. The optimal alternative (e.g. most suitable area or areas for development) is

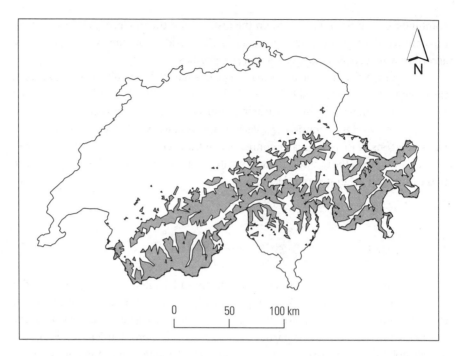

Figure 5.8 Areas in Switzerland with elevation values greater than 2000 m.

that which has the largest score (and thus the highest rank). All that is needed to implement such a system are simple overlay tools and such an approach can readily be applied using vector or raster data (Malczewski, 1999), although most MCDA applications are conducted in the raster environment (see Section 10.2 for a related discussion).

5.4 Case study

This case study makes use of zones (polygons) representing areas in Switzerland with (1) elevation above a threshold value and (2) predicted precipitation amount above a threshold value. Applications of this kind are common in GIS contexts and the results of the analysis could easily be combined with data on other properties such as protected areas or administrative zones. The data on which this analysis is based are described in Section 8.7. Figure 5.8 shows areas with elevations of greater than 2000 m while Figure 5.9 shows areas with a predicted (daily) precipitation amount of greater than 250 1/10 mm (i.e. 250 tenths of a millimetre). In both cases, the vector features were generated from raster grids using the ArcGIS™ software and the edges of the polygon outputs were generalized (smoothed). The aim was to find the common areas and for this purpose the intersect overlay operator was used. The final result (areas that fulfil both criteria) is shown in Figure 5.10.

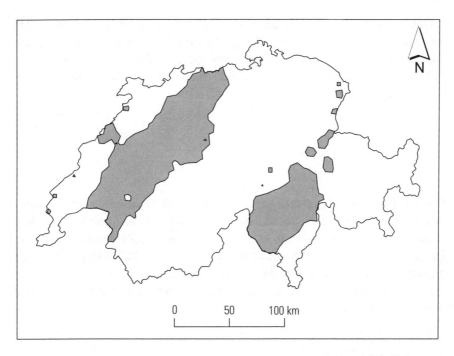

Figure 5.9 Areas in Switzerland with predicted daily precipitation amount values greater than 250 1/10 mm.

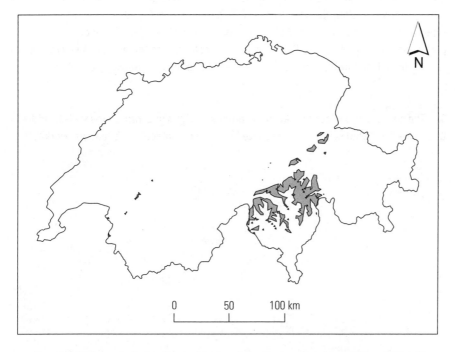

Figure 5.10 Areas in Switzerland with elevation values greater than 2000 m AND daily precipitation amount values of greater than 250 1/10 mm.

Different methods for prediction of precipitation amounts will result in potentially very different outputs (see Chapter 9). Figure 5.10 includes some very small polygons that may not have existed if a different interpolation procedure was used (or the inputs had varied in some other way) and, in such cases, removal of these small polygons might be considered.

Summary

One of the most widely exploited benefits of GIS is their capacity to combine multiple data layers in flexible ways. Such functionality allows complex multiple criteria to be taken into account simultaneously in a way that was practically impossible before the advent of computer-based systems for spatial data analysis. This chapter has provided an overview of some key ways of identifying overlaps between features in different data layers and for combining data layers. In addition, the identification of areas which fulfil multiple criteria was discussed.

Further reading

Overlay operators are described in standard GIS textbooks such as those by Burrough and McDonnell (1998), Heywood *et al.* (2006), Chang (2008), and Longley *et al.* (2005a). Chou (1997), Lee and Wong (2000), and O'Sullivan and Unwin (2002) describe the principles in more detail. Wise (2002) details some key algorithms for overlay. A key reference for GIS-based multicriteria decision analysis is the book by Malczewski (1999).

➡ The next chapter is concerned with network analysis and introduces tools to address questions like 'What is the shortest route between two places on a road network?'.

6

Network analysis

6.1 Introduction

The previous chapter dealt with combining information from different data layers, with one objective being to define connections between objects. This chapter is concerned with connections between places within networks. In particular, methods for the characterization of network complexity and for ascertaining shortest paths between start and end points are detailed. In Section 4.3, approaches for measurement of line lengths were discussed while information on connectivity of vector features was considered in Section 2.2.3. These can be put to use in the analysis of networks. For instance, finding the shortest route between one place and another through a road network is an example of a network analysis approach that necessitates measurement of distance and information on the connections between arcs (representing, for example, roads). Many real-world applications make use of such approaches—the determination of optimal routes for emergency vehicles is one such case. In this chapter some key ways to explore networks are detailed. The account is selective, but will provide a basis on which readers can build.

6.2 Networks

Figure 6.1 shows the constituent parts of a simple network. These include arcs, vertices (representing a change in direction of the arc), and nodes at the end of each arc and connecting different arcs.

Each segment of a network is associated with an impedance factor. In many cases this is simply the distance between nodes, but some alternative factor such as travel cost or time might be used instead. Before network analysis can proceed it is necessary to determine the impedance factors. For a road network, penalties on left turns or right turns (depending on the side of the road on which vehicles drive) might be imposed and u-turns, for example, may or may not be allowed.

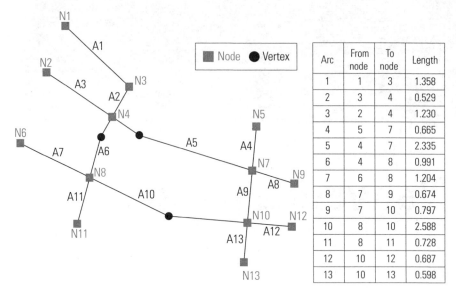

Figure 6.1 Synthetic road network. Arc numbers are prefixed by 'A' and node numbers by 'N'.

The initial focus of this section is on simple summaries of networks, in terms of how connected places are or how complex are the networks which connect these places.

6.3 Network connectivity

The connectivity of a network can be represented using the connectivity matrix (Taafe *et al.*, 1996). The connectivity matrix is a square matrix that contains the arc labels as its column and row headings. The matrix indicates those nodes that are connected by an arc (assigned a value of 1) and those that are not (given 0). The connectivity matrix for the network shown in Figure 6.1 is given in Table 6.1. In this case, nodes 1 and 3, for example, are connected by an arc—that is, the value for column 3 and row 1, or for column 1 and row 3, is 1, indicating a connection.

The matrix in Table 6.1 includes a final column that is the sum of all the row elements and it indicates the number of ways in which a node can be reached in one step from other nodes. The maximum number of possible ways (4) is for nodes 4, 7, 8, and 10. The connectivity matrix, indicated by C^1, in Table 6.1 refers to nodes that are directly connected by arcs—it is termed 'first order'. This idea can be extended to the case of nodes that are connected by two arcs with a further node in between—this is termed 'second order' (and the matrix is indicated by C^2). The matrix in the second-order case is obtained by multiplying C^1 with itself (given by C^1C^1). Matrix multiplication is illustrated below, but further supporting material is given in Appendix A if clarification is required. To multiply a matrix by itself the matrix is multiplied by the transpose (as defined in Appendix A) of itself. Note that, in this case, the result of

Table 6.1 Connectivity matrix for the network shown in Figure 6.1 = matrix $\mathbf{C}^1$

ID	1	2	3	4	5	6	7	8	9	10	11	12	13	Sum
1	0	0	1	0	0	0	0	0	0	0	0	0	0	1
2	0	0	0	1	0	0	0	0	0	0	0	0	0	1
3	1	0	0	1	0	0	0	0	0	0	0	0	0	2
4	0	1	1	0	0	0	1	1	0	0	0	0	0	4
5	0	0	0	0	0	0	1	0	0	0	0	0	0	1
6	0	0	0	0	0	0	0	1	0	0	0	0	0	1
7	0	0	0	1	1	0	0	0	1	1	0	0	0	4
8	0	0	0	1	0	1	0	0	0	1	1	0	0	4
9	0	0	0	0	0	0	1	0	0	0	0	0	0	1
10	0	0	0	0	0	0	1	1	0	0	0	1	1	4
11	0	0	0	0	0	0	0	1	0	0	0	0	0	1
12	0	0	0	0	0	0	0	0	0	1	0	0	0	1
13	0	0	0	0	0	0	0	0	0	1	0	0	0	1

matrix multiplication is the same if the matrix is multiplied by itself and not by its transpose, as the connectivity matrix is symmetric, for example the value for column 2, row 3 is the same as the value for column 3, row 2.

To generate the top left value in the new matrix $\mathbf{C}^2$ take each element in that row (from left to right) from the first matrix (i.e. $\mathbf{C}^1$) and multiply it by each element in that column (from top to bottom) of the second matrix (since we are multiplying one matrix by itself, the first and second matrix are the same). In words, with the column given first and then the row (where ID is the node label given in Table 6.1 and ID(1,1) is 0 while, for example, ID(1,3) is 1):

ID(1,1)×ID(1,1)+

ID(2,1)×ID(1,2)+

ID(3,1)×ID(1,3)+···

and so on until the end of the column and row. At that point add together all the multiplied values. This final summed value is written to cell 1,1 in the new matrix $\mathbf{C}^2$. Note that the row number steps up (increments) in the first matrix while the column number increments in the second.

Next move on to column 2:

ID(1,1)×ID(2,1)+

ID(2,1)×ID(2,2)+

ID(3,1)×ID(2,3)+···

and so on until the end of the column and row. At that point add together all the multiplied values. This final value is written to cell 2,1 in the new matrix $\mathbf{C}^2$.

When results for all columns have been processed, move onto column 1, row 2:

$$\text{ID}(1,2) \times \text{ID}(1,1) +$$

$$\text{ID}(2,2) \times \text{ID}(1,2) +$$

$$\text{ID}(3,2) \times \text{ID}(1,3) + \cdots$$

and so on until the end of the column and row. At that point add together all the multiplied values. This final value is written to cell 2,2 in the new matrix C^2. See Appendix A if further examples of matrix multiplication are required.

This process is completed for all columns and rows. The end result is shown in Table 6.2.

Note that in the case of, for example, connectivity between node 4 and itself, the value of 4 in C^2 indicates movement from node 4 to node 2, 3, 7, or 8 and back to node 4. The matrix for order 3, C^3, is obtained by multiplying the matrices C^1 and C^2 together. The number of meaningful C matrices is determined by the diameter of a network. The diameter of a network is defined as the maximum number of steps needed to move from any to node to any other node in the network using the shortest possible route (Chou, 1997). In the case of the network in Figure 6.1, the network diameter is five—that is, five steps are necessary to move from node 1 to node 12 or node 13. Therefore, in this case a matrix of order 5 is meaningful, but this is not the case for any higher order. Adding the entries of each cell in each of the C matrices (C^1 to C^5 for the example) gives the total accessibility matrix, or T matrix, which indicates the number of ways to move between one node and another in a given number of steps (five in the example) or less (Taafe *et al.*, 1996). Taafe *et al.* (1996) demonstrate the application of the T matrix for identifying the new link that would most markedly increase the

Table 6.2 Connectivity matrix for order 2 = matrix **C^2**

ID	1	2	3	4	5	6	7	8	9	10	11	12	13
1	1	0	0	1	0	0	0	0	0	0	0	0	0
2	0	1	1	0	0	0	1	1	0	0	0	0	0
3	0	1	2	0	0	0	1	1	0	0	0	0	0
4	1	0	0	4	1	1	0	0	1	2	1	0	0
5	0	0	0	1	1	0	0	0	1	1	0	0	0
6	0	0	0	1	0	1	0	0	0	1	1	0	0
7	0	1	1	0	0	0	4	2	0	0	0	1	1
8	0	1	1	0	0	0	2	4	0	0	0	1	1
9	0	0	0	1	1	0	0	0	1	1	0	0	0
10	0	0	0	2	1	1	0	0	1	4	1	0	0
11	0	0	0	1	0	1	0	0	0	1	1	0	0
12	0	0	0	0	0	0	1	1	0	0	0	1	1
13	0	0	0	0	0	0	1	1	0	0	0	1	1

connectivity of a road network. Tools like the T matrix are potentially very useful aids to transport network development. The following section details some single-value summaries of network characteristics.

6.4 Summaries of network characteristics

Several simple summaries of network characteristics exist. These include the γ (gamma) index and the α (alpha) index (Taafe *et al.*, 1996), both of which are introduced below. A better connected network has larger values of γ and α (Chou, 1997) and together they provide a useful summary of network complexity and connectivity.

For a given number of nodes, more arcs indicate greater connectedness. The minimum number of arcs (or links), l, needed to connect n nodes is given by:

$$l_{min} = n - 1 \tag{6.1}$$

With a minimally connected network, removal of any one arc will result in two unconnected networks—that is, there are no loops or circuits in the network. For the network in Figure 6.1, the number of nodes is the same as the number of arcs—there is a circuit comprising arcs 5, 6, 10, and 9. It is therefore not minimally connected, but it would be if arc 5 or arc 10 was removed from the circuit.

There is a range of measures of network characteristics, such as complexity of a network or the degree to which it is connected. The γ index is one summary of network complexity. It is the ratio of the number of links (arcs) in a network to the maximum number of links possible. For a planar graph with n nodes, the maximum number of links is given by $3(n-2)$. A graph is a set of nodes connected by arcs; a planar graph has no intersecting arcs—the intersections in Figure 6.1 are represented by nodes which connect separate arcs and so it is a planar graph. In non-planar graphs, such as three-dimensional air transport networks, the maximum number of links is $n(n-1)/2$ (Chou, 1997). The γ index for a planar graph is given by (Chou, 1997):

$$\gamma = \frac{l}{l_{max}} = \frac{l}{3(n-2)} \tag{6.2}$$

where l is the number of links in the network and l_{max} is the maximum number of possible links (i.e. $3(n-2)$). The γ index can take values from 0 to 1 where small values indicate simpler networks with few links and larger values indicate more links, and therefore a better connected network (Chou, 1997).

The network illustrated in Figure 6.1 has 13 nodes and 13 lines (arcs), so $n=13$ and $l=13$. In this case, γ is calculated by:

$$\gamma = \frac{13}{3(13-2)} = \frac{13}{33} = 0.394$$

If five links were added to the existing set of links and nodes (the number of nodes remaining the same, e.g. adding a link between the existing nodes 3 and 7) the updated calculations would be:

$$\gamma = \frac{18}{3(13-2)} = \frac{18}{33} = 0.545$$

The larger value of γ than in the previous case reflects the larger number of connections.

An additional measure of network connectivity measures the number of circuits, c, that exist within a network. A circuit has a start node that is the same as the end node and it comprises a closed loop (Lee and Wong, 2000). In a minimally connected network (defined previously in this section) there are no circuits and the number of circuits can be calculated by subtracting the number of arcs required to form a minimally connected network from the observed number of arcs in the network. This is given by $l-n+1$ (recall that l is the number of arcs and n is the number of nodes). In the case of the network shown in Figure 6.1, $l=13$, which gives $13-13+1=1$, i.e. there is one circuit. The α index is the ratio of the number of circuits (c) to the maximum possible number of circuits in a network (c_{max}) (Chou, 1997). It is given by:

$$\alpha = \frac{c}{c_{max}} = \frac{l-n+1}{2n-5} \tag{6.3}$$

For the network in Figure 6.1, this gives $^1/_{21} = 0.048$.

Scott *et al.* (2006) review the γ index, amongst other measures, and present an additional measure for assessing the importance of particular links and for evaluating the connectivity of networks. Such measures may be useful, for example, in informing the design of new road networks or in altering those already in existence.

6.5 Identifying shortest paths

Identification of the shortest path between two places is a common aim in the analysis of networks. A widely used approach to identifying the shortest path between two places is the algorithm presented by Dijkstra (1959). The basis of this approach is that links, connected to a starting node, are selected that have the shortest path back to that starting node, and the algorithm is outlined below. Such an approach is much more efficient than an approach that simply calculates all possible routes and selects the shortest one. A worked example is given based on the network shown in Figure 6.2.

In the example, the objective is to identify the shortest path from node 1 to node 6. This is a straightforward task in this case, but in most real-world cases some automated procedure is necessary. With the initial step, represented using the 'Starting

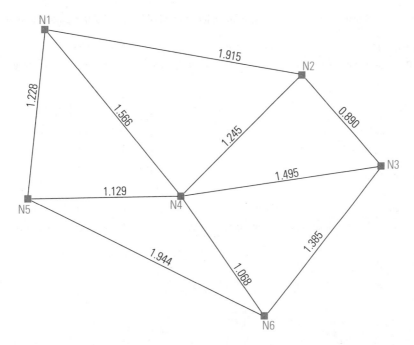

Figure 6.2 Network used for illustrating the shortest path algorithm, with the length of each arc indicated. 'N' indicates node.

table' in Table 6.3, each node has a distance from the source node that is infinity (∞) and the next node back from each node (parent) to the origin is blank. 'Included' indicates nodes which have comprised part of the shortest route to a node at any stage. Note that, using this approach, costs (e.g. distances) must be positive. The example is presented numerically, rather than graphically, as a key concern is to show exactly how the algorithm operates.

The description of the shortest path algorithm that follows is based on that given by Wise (2002) and makes reference to the format shown in Table 6.3:

- Select the non-included node with the shortest distance to the source.
- Include this node (i.e. replace 'N' in the final column with 'Y'). The distance from this node back to the source node is indicated by dist(n).
- Identify the nodes connected to node n that are still not included (they have an entry 'N' in their final column). The distance between each of these nodes (m) and the node n is given by d(nm).
- For each of the m nodes:
  if dist(n) + d(nm) < dist(m) then dist(m) = dist(n) + d(nm) and parent(m) = n
  that is, if there is a shorter route from the source node to node m via node n then the entry for node m is updated with the new total distance figure and the parent value for node m is set to n (the previous node along the path from the origin node). The algorithm is illustrated below with reference to the sub-tables that comprise Table 6.3.

Table 6.3 Starting table and table after each iteration of Dijkstra's shortest path algorithm

Starting table			
N	Distance	Parent	Included
1	∞		N
2	∞		N
3	∞		N
4	∞		N
5	∞		N
6	∞		N

After iteration 1			
N	Distance	Parent	Included
1	0	–	Y
2	1.915	1	N
3	∞		N
4	1.566	1	N
5	1.228	1	N
6	∞		N

After iteration 2			
N	Distance	Parent	Included
1	0	–	Y
2	1.915	1	N
3	∞		N
4	1.566	1	N
5	1.228	1	Y
6	3.172	5	N

After iteration 3			
N	Distance	Parent	Included
1	0	–	Y
2	1.915	1	N
3	3.061	4	N
4	1.566	1	Y
5	1.228	1	Y
6	2.634	4	N

After iteration 4			
N	Distance	Parent	Included
1	0	–	Y
2	1.915	1	Y
3	2.805	2	N
4	1.566	1	Y
5	1.228	1	Y
6	2.634	4	N

After iteration 5			
N	Distance	Parent	Included
1	0	–	Y
2	1.915	1	Y
3	2.805	2	N
4	1.566	1	Y
5	1.228	1	Y
6	2.634	4	Y

Firstly, the distance from node 1 is set to 0, as it is the source node. The 'Included' entry for node 1 is set to Y, to indicate it is part of the shortest path (obviously it must be as it is the source node!). Node 1 is connected to nodes 2, 4, and 5, and the distances from node 1 to these nodes are smaller than infinity and so replace the existing entries in the distance column. The parent nodes for nodes 2, 4, and 5 are set to 1. The result is the table 'After iteration 1' (where each iteration represents a single cycle in the process).

The shortest distance for any node that has not yet been included is that for node 5 (a distance of 1.228) and this node is now included. Node 5 is connected to two nodes down the path from node 1: nodes 4 and 6. The distance to node 4 via node 5 is larger than the distance already recorded for node 4, which is direct from node 1, so the entry for node 4 remains the same. The distance from node 5 to node 6 is 1.944 units. This is added to the distance from node 1 to node 5 (1.228), giving a total path length from node 1 via node 5 to node 6 of 3.172 and the parent node for node 6 is set to 5 (i.e. the connection back to node 1 is via node 5). This distance replaces the infinity value for node 6. This gives the table 'After iteration 2'.

The shortest distance for any non-included node is 1.566, for node 4, so this node is now included. Node 4 is connected to nodes 2, 3, and 6. The distance from node 1 to node 2 via node 4 is greater than the direct link between node 1 and node 2, so the entry for node 2 remains the same. The distance from node 1 to node 3 via node 4 is 1.566 + 1.495 = 3.061 and this value replaces the infinity value for node 3, for which the parent node is set to 4. The distance from node 1 to node 6 via node 4 is 1.566 + 1.068 = 2.634. This is smaller than the existing distance value for node 6, so the distance value is replaced and the parent node is changed to 4. The result is the table 'After iteration 3'.

The shortest distance for any node that has not yet been included as part of the shortest route is 1.915. This is the distance value for node 2 and node 2 is included at this stage. Node 2 is connected to nodes 3 and 4. The distance from node 1 to node 3 via node 2 is $1.915+0.890=2.805$ and is smaller than the existing distance value for node 3, so the distance value for node 3 is replaced and the parent node changed to 2. Node 4 is already included so it is not considered. The end result is the table 'After iteration 4'.

The distance of 2.634, for node 6, is the smallest distance for a non-included node. Node 6 is included at this stage. Node 6 is connected to the remaining non-included node, node 3. The distance from node 1 to node 3 via nodes 4 and 6, however, is greater than the existing distance value so it remains the same. This is the final stage and results in the table 'After iteration 5'.

The distance values for each node are the shortest path distances to that node from node 1. For example, the shortest path from node 1 to node 4 is 1.566 units while the shortest path from node 1 to node 6 is via node 4 and is 2.634 units $(1.566+1.068)$. Wise (2002) and Chang (2008) provide further worked examples of the use of the shortest path algorithm.

6.6 The travelling salesperson problem

There are many other methods for the analysis of networks that could be outlined. For instance, algorithms for solving the so-called 'travelling salesperson problem' (TSP) are often encountered in GIS textbooks. The TSP entails finding the shortest possible route that visits each of a set of points once only and returns to the starting point. The number of possible routes increases markedly with the number of locations that must be visited. Where there are n locations that must be visited, including the origin (e.g. the depot), the number of possible routes is given by $(n-1)!$, where ! indicates the factorial of a number. As an example, 3! is $3\times2\times1=6$. In practice, most trips can be taken in two directions and then the possible number of routes is given by $(n-1)!/2$ (Longley et al., 2005a), so if there are eight locations to be visited, $n-1=7$ and $7!=7\times6\times5\times4\times3\times2\times1=5040$ and the number of possible routes is $5040/2=2520$. In practice, for a large number of places to visit, it is not feasible to identify the optimal route, but approaches exist to rapidly identify routes that may not be the best, but which approach the optimal route according to some criterion. Wise (2002) provides a summary of the TSP and possible approaches to it.

6.7 Location–allocation problems

A major class of problems for which network analysis tools provide a solution is the matching of resources to particular people or groups. The term 'location–allocation' refers to the location of facilities and the allocation of resources from a given facility to particular locations. For example, we may wish to determine which depot may make deliveries of goods to which particular areas within a maximum travel time limit (using information on road distances and perhaps other impedance factors such as average travel time). Information on supply and demand levels may be used to determine the maximum catchment area of a facility (see Birkin *et al.* (1996) for a summary). Approaches to solving the location–allocation problem may be used to help ascertain if a service provider can meet the needs of a given area or if a new facility (e.g. a hospital or a retail outlet) might be needed or existing facilities expanded. Chang (2008) provides an introduction to location–allocation modelling with several examples.

6.8 Case study

As in most other chapters, this chapter concludes with a case study that can be conducted using standard GIS software. The data are provided on the book website so that

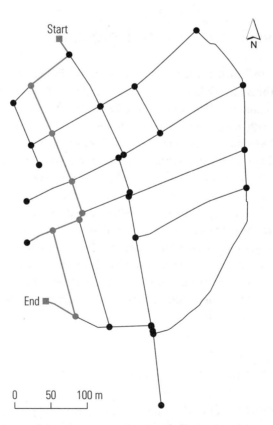

Figure 6.3 Shortest path between start and end point. The nodes of the network are given by circles and the start and end points by squares.

readers can explore the application of the selected methods. Guidance notes are also provided on the website to outline how the methods are applied in these particular cases.

The data used in this case study represent the street layout of the town of Aberystwyth, captured from historic maps as part of a project exploring the morphology (i.e. the shape) of various medieval town plans in England and Wales[1] (Lilley *et al.*, 2007). In this case study, the objective was to identify the shortest path between the start and end points indicated in Figure 6.3. Such problems are common, for example any road user may wish to identify the shortest route between their starting point and their destination. Information on toll costs, traffic density, and other issues that may affect the choice of route can easily be taken into account. ArcGIS™ Utility Network Analyst offers a routine to compute shortest paths along networks and this software was used to identify the shortest path, which is indicated by a heavy line in Figure 6.3.

With an efficient approach like Dijkstra's algorithm, it is possible to rapidly compute shortest paths while taking into account factors such as traffic jams where such information is available.

Summary

This chapter introduced a variety of approaches to characterizing networks which may have application in, for example, transportation planning contexts. An algorithm for identifying shortest paths was also outlined. Brief summaries of some important network analysis problems were also provided. This chapter provides a foundation for understanding a major class of approaches for the analysis of vector data, and it presents the basic principles that are then foundations of a range of methods frequently applied to solve real-world problems. The suggested further reading provides a way forward if these kinds of approaches are of interest.

Further reading

Wise (2002) details some algorithms for network analysis. That book provides a good starting point for readers wanting to know more about GIS algorithms as a whole. Further accounts of network analysis are provided by Chou (1997), Lee and Wong (2000), and Chang (2008). Taaffe *et al.* (1996) provide an in-depth account of transportation geography that provides the context for the material covered in this chapter.

⮕ This chapter was concerned with line features and their analysis; the analysis of point patterns (i.e. sets of point 'events') is the subject of the following chapter.

1 http://www.qub.ac.uk/urban_mapping/

7

Exploring spatial point patterns

7.1 Introduction

The simplest form of spatial data is a spatial point pattern, although approaches to their analysis are no simpler than is the case with data that have attributes (i.e. have numeric values attached to them). A spatial point pattern is simply a set of locations that correspond to events. For example, a set of points that shows the locations of trees or of people with a particular disease is a point pattern. Many tools exist for analysing point patterns. Such approaches allow assessment of, for example, the degree to which point events are clustered or dispersed. Synthetic examples of clustered and dispersed point patterns are given in Figure 7.1 (hereafter referred to as 'Point Pattern 1' or 'PP1') and Figure 7.2 ('PP2'). The concern is often to consider if the point pattern is spatially structured in some particular way—if the points are clustered then this suggests that events are more likely to occur in some places than in others. Conversely,

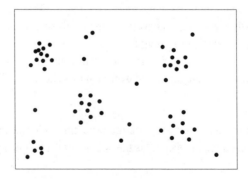

Figure 7.1 Clustered point pattern (PP1).

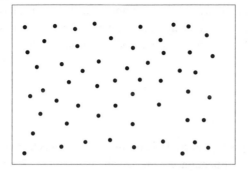

Figure 7.2 Dispersed point pattern (PP2).

if the points are dispersed, this suggests that events are likely to occur as far away as possible from other events. This chapter introduces a variety of ways for assessing and characterizing spatial point patterns. A set of points that also have values attached to them is called a marked point pattern, but the concern in this chapter is with points that have no attached attribute data. While much of the early development of approaches was in an ecological context, applications areas are now extensive. Typical applications of point pattern analysis include the exploration of clustering in disease events (an interesting study by Openshaw *et al.* (1993) notes that cases of some forms of cancer tend to cluster while others do not) and clustering in particular species of tree (relevant references appear at the end of the chapter).

Point pattern analysis can be divided into two sets of approaches: those that deal with first-order effects and those that deal with second-order effects. First-order effects are referred to in terms of intensity—that is, the mean number of events per unit area at a given location. Second-order effects, or spatial dependence, refer to the relationship between paired events in the study region (Bailey and Gatrell, 1995). So, the first refers to the number of events in an area while the second refers to structure in the point pattern. As an example of the latter, if the number of events in areas separated by a fixed distance is consistently similar for all locations and similarity in the number of events in two areas decreases as distance between these locations increases then there is spatial dependence in the point pattern at a variety of scales. First-order effects are considered in Section 7.3, while second-order effects are the subject of Section 7.4. First- and second-order effects are a function of spatial scale. In the former case the mean intensity may change smoothly from place to place over a large area. In the latter case, features with a finer scale are the concern (Bailey and Gatrell, 1995). In practical terms, it is often difficult to separate first- and second-order effects (O'Sullivan and Unwin, 2002). This chapter does not delve further into the particular problem of distinguishing between the two, but simply presents a set of tools for the analysis of point patterns and provides some pointers for their use.

Before proceeding, it is worth making one comment of note. If, for example, a point pattern represents disease events, there may be clusters of events in some places simply because there are more people in an area, for example an urban area. If we want to explore spatial clustering, it is necessary to account for the total population, often referred to as the 'population at risk'. Methods exist that control for the population at risk but this chapter focuses on the standard unmodified measures. Nevertheless, it is an issue that should be taken into account when considering the approaches presented.

7.2 Basic measures

Analysis of spatial point patterns is, like any analysis of any form of spatial data, likely to begin with visual inspection. Again, as with any form of spatial analysis, a second step may be to compute one of a variety of descriptive measures. Two quite widely used summary measures are the mean centre and the standard distance. The mean centre of a point pattern indicates the central tendency of the points. It is simply the mean of the x and y coordinates:

$$\bar{\mathbf{x}} = (\mu_x, \mu_y) = \left(\frac{\sum_{i=1}^{n} x_i}{n}, \frac{\sum_{i=1}^{n} y_i}{n} \right) \tag{7.1}$$

The bold letter x indicates a vector representing location (i.e. x and y coordinates) and the bar above it indicates that it is the mean average value: μ_x is the mean of the x values and μ_x is the mean of the y values. The number of events is given by n. Dispersion around the mean centre can be measured with the standard distance, d_s:

$$d_s = \sqrt{\frac{\sum_{i=1}^{n}(x_i - \mu_x)^2 + (y_i - \mu_y)^2}{n}} \tag{7.2}$$

The mean centres and standard distances for the two point patterns illustrated in Figures 7.1 (PP1) and 7.2 (PP2) are given in Figures 7.3 and 7.4.

The mean centres and standard distances for the two examples are similar with respect to the common boundary box. However, the characteristics of the two point patterns are very different in other ways. For example, the intensity of points in Figure 7.1 is

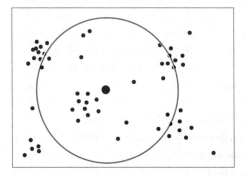

Figure 7.3 PP1: mean centre (large point) and standard distance (circle).

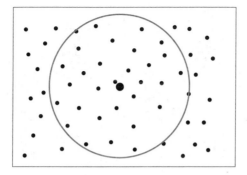

Figure 7.4 PP2: mean centre (large point) and standard distance (circle).

variable across the study region (as noted above, it is a clustered point pattern) while the intensity of points in Figure 7.2 does not vary markedly from place to place, it is a fairly regular point pattern. The rest of this chapter will focus on methods for exploring the degree to which a point pattern is clustered either globally (i.e. across the whole study area) or locally (i.e. in regions within the whole study area).

7.3 Exploring spatial variations in point intensity

A key concern in point pattern analysis is exploration of spatial variation in point pattern intensity. The following sections explore some ways of mapping event intensity. Such methods are used for exploring first-order effects (as defined in Section 7.1). The first focus is on quadrat analysis and the second focus is on a more sophisticated means of assessing spatial variation in event intensity, kernel estimation.

7.3.1 Quadrats

A common way of exploring spatial patterning in the number of events is quadrat analysis. At its simplest level, this entails superimposing a regular grid over the point pattern and counting how many points fall within each grid square. This is illustrated in Figures 7.5 and 7.6 for the point patterns shown earlier. Clearly, varying quadrat size will have an impact on analyses, as will aggregation at different levels in other applications (recall that Section 4.9 dealt with such effects). In addition, the origin of the quadrat grid (its smallest x and y coordinates) will also influence the results.

In Figure 7.5 (PP1), there are several counts that are greater than 2. In contrast, in Figure 7.6 (PP2), there are no counts greater than 2. These values reflect the clustering and regularity, respectively, in the two point patterns. There is a range of ways of assessing the degree of clustering or regularity. One possible approach is to use a measure of spatial autocorrelation (such as Moran's I, as defined in Section 4.8) to assess the degree of spatial dependence in quadrat counts. For rook's case contiguity (see Section 4.8), I for the quadrat counts in Figure 7.5 (with zeros placed in the empty cells) was 0.0017

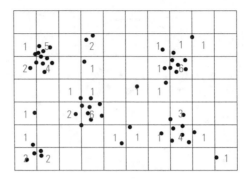

Figure 7.5 PP1: quadrats.

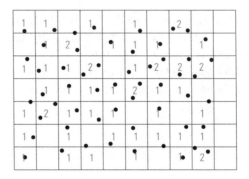

Figure 7.6 PP2: quadrats.

while for queen's case (again, see Section 4.8) it was 0.0887. For rook's case contiguity, *I* for the quadrat counts in Figure 7.6 was −0.0395 while for queen's case it was −0.1053. For the quadrats used in the example, PP1 appears clustered, i.e. it is positively spatially autocorrelated (albeit not to a marked degree), while PP2 appears regular or dispersed, i.e. it is negatively autocorrelated, thus confirming the impression gained from visual inspection. The *I* coefficient was computed using the GeoDa software (Anselin *et al.*, 2006). The value of *I* is, of course, partly a function of the size of the quadrats.

A simple means of assessing the degree of clustering or regularity in a point pattern is the variance/mean ratio (VMR). The VMR indicates the degree to which a point pattern departs from that predicted by the Poisson distribution, which is often used to express the probability of events occurring in a given area. The Poisson distribution can be given by:

$$P(k) = \frac{\lambda^k e^{-\lambda}}{k!}$$

(7.3)

where λ is the mean average intensity of the point pattern, k is the number of events, and e is the base of the natural logarithm ($=2.71828...$). In words, the equation gives the predicted fraction of quadrats containing k events for a mean intensity λ (O'Sullivan and Unwin (2002) provide another account). In Figures 7.5 and 7.6 there are 70 quadrats and 55 events. In both cases, the mean quadrat count (the mean intensity) is $55/70 = 0.786$. As an example, the probability of there being three events in a quadrat is given by:

$$P(3) = \frac{0.786^3 e^{-0.786}}{3!} = \frac{0.485588 \times 0.455664}{3 \times 2 \times 1} = \frac{0.221265}{6} = 0.036877$$

Events like the number of points in quadrats can be modelled using the Poisson distribution and a point pattern with a Poisson distribution may be described as confirming to complete spatial randomness (CSR)—that is, there is no apparent structure. Table 7.1 gives quadrat counts (given by #) for PP1 and PP2, the counts expressed as proportions and $P(k)$. Visual inspection of the two sets of fractions suggests that the fractions for PP2 are much closer to those predicted given the Poisson distribution than are those for PP1.

Following on from the discussion above, the VMR is expected to have a value of 1 if the distribution is Poisson—a value greater than 1 indicates clustering while a value of less than 1 indicates a dispersed point pattern. Recall that, for the examples in Figures 7.5 and 7.6, there are 70 quadrats and 55 events, giving a mean quadrat count of $55/70 = 0.786$ in both cases. For the point pattern in Figure 7.5 (PP1), the variance (i.e. the mean squared difference between each quadrat count and the mean quadrat count) is $131.786/70 = 1.883$. For the point pattern in Figure 7.6 (PP2), the variance is $29.786/70 = 0.426$. The variance/mean ratios are then given by:

$$\frac{1.883}{0.786} = 2.396$$

$$\frac{0.426}{0.786} = 0.542$$

Table 7.1 Quadrat counts for PP1 and PP2, fractions and $P(k)$

k	PP1#	PP2#	PP1 fraction	PP2 fraction	P(k)
0	42	24	0.600	0.343	0.45566381
1	17	37	0.243	0.529	0.35815175
2	5	9	0.071	0.129	0.14075364
3	1	0	0.014	0.000	0.03687745
4	2	0	0.029	0.000	0.00724642
5	1	0	0.014	0.000	0.00113914
6	2	0	0.029	0.000	0.00014923
7	0	0	0.000	0.000	0.00001676
8	0	0	0.000	0.000	0.00000165
9	0	0	0.000	0.000	0.00000014
10	0	0	0.000	0.000	0.00000001

This supports the notion of PP1 exhibiting clustering (as the value is greater than 1), while visual inspection of PP2 suggests dispersal (with a value of less than 1). This supports the impression gained from inspection of Table 7.1. In addition, the analysis using Moran's I coefficient, summarized above, reached similar conclusions. Yet another way of assessing the degree to which a point pattern conforms to CSR is to conduct a chi-square (χ^2) test (O'Sullivan and Unwin, 2002). This is given by the variance (without division by n) of the quadrat counts divided by the mean:

$$\chi^2 = \frac{\sum (k - \mu_k)^2}{\mu_k} \tag{7.4}$$

where μ_k is the mean quadrat count (the estimate of λ). The calculations for the variances of PP1 and PP2 are given in Table 7.2.

Note that k and μ are the same for PP1 and PP2. The χ^2 statistic for PP1 and PP2 is then calculated:

$$\text{PP1: } \chi^2 = \frac{131.786}{0.786} = 167.666$$

$$\text{PP2: } \chi^2 = \frac{29.786}{0.786} = 37.895$$

The number of degrees of freedom (see Section 3.2) is given by the number of quadrats (70) minus 1, i.e. 69. Examination of a table of critical values of χ^2 (e.g. see Ebdon (1985), Appendix C) indicates that the value for PP1 is significant to greater than the 0.001 level, while the value for PP2 is not significant even at the 0.1 level. In other words, in the case of PP1 we can reject the null hypothesis that the point pattern was generated by CSR. In the case of PP2 we cannot reject the null hypothesis (see Section 3.4 for a discussion on a related topic).

The following section builds on the account of geographical weighting schemes in Section 4.7, but in a point pattern analysis context.

Table 7.2 Quadrat counts for PP1 and PP2: derivation of the variances

k	PP1#	PP2#	PP1 and PP2 $(k-\mu)$	PP1 and PP2 $(k-\mu)^2$	PP1 $\#(k-\mu)^2$	PP2 $\#(k-\mu)^2$
0	42	24	−0.786	0.617796	25.947432	14.827104
1	17	37	0.214	0.045796	0.778532	1.694452
2	5	9	1.214	1.473796	7.368980	13.264164
3	1	0	2.214	4.901796	4.901796	0.000000
4	2	0	3.214	10.329796	20.659592	0.000000
5	1	0	4.214	17.757796	17.757796	0.000000
6	2	0	5.214	27.185796	54.371592	0.000000
Total	70	70			131.785720	29.785720

7.3.2 Kernel estimation

Kernel estimation (KE) is a more sophisticated means of exploring spatial variations in event intensity. Such approaches are often used in the identification of clustering 'hot spots'. If we want to estimate the intensity of points over an area we can simply calculate the number of events within a radius around the nodes of a grid and divide this amount by the area concerned. In other words, we in effect overlay onto the point pattern a set of points on a regular grid and compute the event intensity within the neighbourhood of each point on the grid. This is called the naïve estimator. The naïve intensity estimate is given by:

$$\hat{\lambda}(\mathbf{x}) = \frac{\#(C(\mathbf{x},d))}{\pi d^2} \tag{7.5}$$

where $\#(C(\mathbf{x},d))$ indicates the number (#) of events in the circle $C(\mathbf{x},d)$ that has as its centre the location $\mathbf{x}$, and has the radius d, with its area given by πd^2. The bold lower case $\mathbf{x}$ is matrix vector notation as introduced in Section 3.3. In this case, $\mathbf{x}$ refers to a location with coordinates x, y.

Figure 7.7 shows intensity estimates for PP1 using radii of 25 and 50 units. The differences in minimum and maximum intensities and in the spatial patterning are immediately apparent. The maximum intensity is greater for a radius of 25 units as this smaller radius 'picks out' the small clusters. Their effect is diminished when the larger radius (50 units) is used. The two groups of clusters notable when the 25 unit radius is used become, in effect, merged into one as the size of the radius is increased to 50 units.

KE can be expanded to make use of a geographical weighting scheme (a kernel function) whereby the influence of the points varies depending on how far they are from the centre of the window. The general idea of geographical weighting was

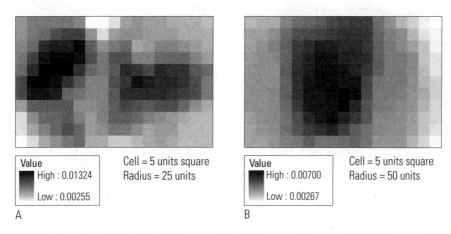

Value		Cell = 5 units square
High : 0.01324		Radius = 25 units
Low : 0.00255		

A

Value		Cell = 5 units square
High : 0.00700		Radius = 50 units
Low : 0.00267		

B

Figure 7.7 Intensity estimates for PP1 using radii of 25 units (A) and 50 units (B). The values are events per square unit.

explored in Section 4.7, but is recapped here to illustrate KE. The KE of intensity is given by:

$$\hat{\lambda}_k(\mathbf{x}) = \sum_{i=1}^{n} \frac{1}{\tau^2} k\left(\frac{\mathbf{x} - \mathbf{x}_i}{\tau}\right) \tag{7.6}$$

where τ is the bandwidth (determining the size of the kernel, see Section 8.4 for a related discussion) and $\mathbf{x} - \mathbf{x}_i$ indicates the distance between the centre of the kernel ($\mathbf{x}$) and the location $\mathbf{x}_i$. A kernel function is used to give larger weights to nearby events (points) than events that are more distant.

There is a variety of different kernels that have been used for KE. The quartic kernel is encountered frequently in the point pattern analysis literature (Bailey and Gatrell, 1995):

$$k(\mathbf{u}) = \begin{cases} \dfrac{3}{\pi}(1 - \mathbf{u}^T\mathbf{u})^2 & \text{for } \mathbf{u}^T\mathbf{u} \leq 1 \\ 0 & \text{otherwise} \end{cases} \tag{7.7}$$

where $\mathbf{u}$ is d_i/τ and d_i is the distance from the centre of the kernel, and superscript T indicates the transpose of the matrix (see Appendix A for a definition). In short, values are summed for all cases where the distance is less than or equal to the bandwidth. The quartic kernel is illustrated in Figure 7.8 and a section through the quartic kernel is shown in Figure 7.9. The KE with the quartic kernel can be given by (Bailey and Gatrell, 1995):

$$\hat{\lambda}_k(\mathbf{x}) = \sum_{d_i \leq \tau} \frac{3}{\pi\tau^2}\left(1 - \frac{d_i^2}{\tau^2}\right)^2 \tag{7.8}$$

The application of this equation is outlined below.

The naïve estimator and a kernel estimator are illustrated with reference to two sets of points, labelled A and B, in Figure 7.10. We will take each set as a subset (contained

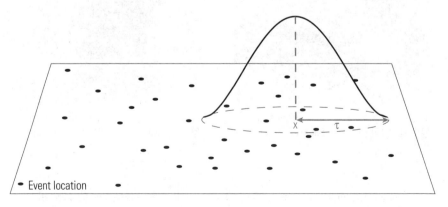

Figure 7.8 Quartic weighting scheme, with bandwidth τ.

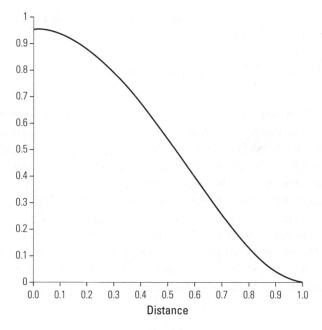

Figure 7.9 Quartic kernel section.

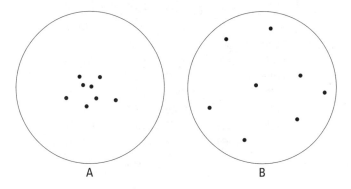

Figure 7.10 Points within a moving window with a radius of 25 units.

within the circles given, which represent a moving window with a radius of 25 units) of a larger point pattern. In both cases, there are eight points.

Using the naïve estimator, both point patterns A and B have the same intensity—the area of the circle is given by $\pi\tau^2 = 3.14159 \times 25^2 = 1963.495$. Given Equation 7.5, in both cases the intensity estimate is given by $8/1963.495 = 0.004074$. Using a kernel estimator, point pattern A has a greater intensity than point pattern B.

Kernel estimation is now illustrated using the same example. Working through Equation 7.8, the left-hand side (LHS) of the equation, $3/\pi\tau^2$, is computed first. The bandwidth, τ, is set at 25 units, so $3/\pi\tau^2 = 3/(3.14159 \times 25^2) = 0.00152$.

The right-hand side (RHS) of Equation 7.8 is $(1 - d_i^2/\tau^2)^2$. Taking one distance value as an example (0.927) this leads to:

$$\left(1 - \frac{0.927^2}{25^2}\right)^2 = \left(1 - \frac{0.85933}{625}\right)^2 = 0.99863^2 = 0.99726$$

Tables 7.3 (for point pattern A) and 7.4 (for point pattern B) show the x and y coordinates of each event, the distance of the event from the window centre, and computed values of the RHS of Equation 7.8 followed by the LHS of the equation (which has a value of 0.00152 for all cases given a bandwidth of 25 units) multiplied by the RHS. The sum of products in this last column is also given.

The difference in the estimated intensities for the two methods will decrease as the size of the bandwidth increases. This is indicated by Figure 7.11, which shows the KE values for bandwidths of 5, 10, 15, 20, 25, and 30 units. Note how for point pattern A, which is visibly more clustered around the centre of the window, the KE value is much larger for a bandwidth of five units than is the value for point pattern B, but this difference decreases as the size of the kernel bandwidth is increased.

Table 7.3 Point pattern A: intensity estimate using the quartic kernel with a bandwidth of 25 units (distance is from the centre of the kernel)

x	y	Distance	RHS	LHS×RHS
−16.670	−2.156	0.927	0.997	0.00152
−19.794	−1.079	2.378	0.982	0.00150
−15.377	−5.496	4.235	0.943	0.00144
−14.084	1.507	4.832	0.927	0.00142
−21.087	1.615	4.960	0.923	0.00141
−18.717	−8.082	6.330	0.876	0.00134
−25.613	−5.389	8.802	0.767	0.00117
−8.912	−6.035	9.592	0.727	0.00111
			Sum	0.01091

Table 7.4 Point pattern B: intensity estimate using the quartic kernel with a bandwidth of 25 units (distance is from the centre of the kernel)

x	y	Distance	RHS	LHS×RHS
−19.687	−1.079	2.276	0.983	0.00150
−4.603	2.153	13.554	0.499	0.00076
−5.572	−12.392	15.948	0.352	0.00054
−23.565	−19.180	18.334	0.214	0.00033
−35.309	−8.513	18.966	0.180	0.00028
−14.731	17.776	19.838	0.137	0.00021
−29.815	14.220	20.225	0.119	0.00018
3.801	−3.449	21.409	0.071	0.00011
			Sum	0.00390

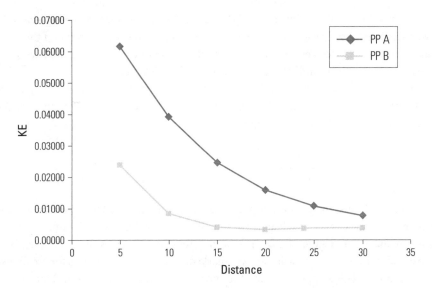

Figure 7.11 KE for point patterns A and B for different bandwidths.

Clearly, the size of the kernel bandwidth will have a marked impact on the intensity estimates. In practice, a variety of bandwidths is often used and the difference in results assessed. In the cases described above, the size of the kernel bandwidth is the same at all locations, but techniques exist such that the bandwidth can be varied from place to place as a function of local event intensity, with a small bandwidth in areas with intense point patterns and a large bandwidth in areas with a low event intensity (Bailey and Gatrell, 1995; Brunsdon, 1995).

7.4 Measures based on distances between events

The second-order properties of a point pattern (as defined in Section 7.1) can be measured using information on distances between events. This section deals with such approaches. The initial concern is with methods that make use of information on the nearest neighbours of events. The second concern is with the K function, a widely used means of assessing the degree of spatial dependence in point patterns (i.e. does the number of points in one area tend to be similar to the number of points in another area some fixed distance away?)

7.4.1 Nearest-neighbour methods

Various measures exist that are based on the nearest neighbour to each event. The mean nearest-neighbour distance is one such measure and it is given by:

$$\bar{d}_{\min} = \frac{\sum_{i=1}^{n} d_{\min}(\mathbf{x}_i)}{n} \tag{7.9}$$

where n is the number of events and d_{min} is the distance from the nearest event. In words, the distance of each event i at location $\mathbf{x}_i$ to its nearest neighbour and the average of these distances are computed. These statistics are used to compute two other measures called the G and F functions, which will be described next.

The G and F functions allow the exploration of event to event nearest-neighbour distances. The G function is defined as the cumulative frequency distribution of the nearest-neighbour distances. It is given by:

$$G(d) = \frac{\#(d_{min}(\mathbf{x}_i) < d)}{n} \tag{7.10}$$

In words, $G(d)$ gives the proportion (since the count is divided by n) of nearest-neighbour distances that are less than distance d. Obtaining G for different values of d enables assessment of the degree of clustering at different spatial scales. For example, for a clustered point pattern, G will increase markedly as distance increases for small distances. For a regular point pattern the increase will be more gradual. Figure 7.12 shows the G function for the point patterns in Figures 7.1 (PP1) and 7.2 (PP2) for distance steps of 2.5 units. Clearly, the values of G for PP1 are larger at smaller distances than they are for PP2, indicating the greater degree of clustering in PP1 than in PP2. A value of 1 for G corresponds to distances that are larger than the maximum nearest-neighbour distance for a given point pattern. In other words, for PP1, no nearest-neighbour distance is greater than 15 units, while for PP2 no nearest-neighbour distance is greater than 12.5 units.

The F function is similar to the G function, but instead of the events a sample of point locations is selected randomly from anywhere in the study area—that is,

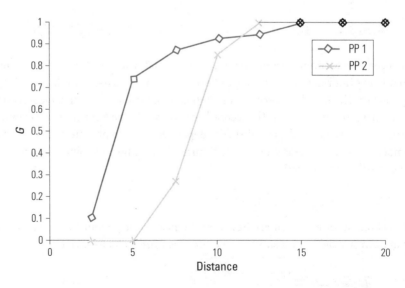

Figure 7.12 G function for the point patterns in Figures 7.1 (PP1) and 7.2 (PP2).

the nearest-neighbour distances are computed for randomly selected locations and not for the point event locations. The F function is given by:

$$F(d) = \frac{\#(d_{\min}(\mathbf{x}_i, X) < d)}{m} \tag{7.11}$$

where $d_{\min}(\mathbf{x}_i, X)$ is the minimum distance from the location $\mathbf{x}_i$ in the randomly selected set of locations to the nearest event in the point pattern, X, and m is the number of randomly selected locations. Both G and F can be plotted against d to allow exploration of changes in clustering with distance. The G and F functions may vary, for example, for a clustered pattern, values of G may be larger than values of F for small distances as in the latter case only a random sample is used and the impact of clusters may be diminished (Bailey and Gatrell, 1995; O'Sullivan and Unwin, 2002). One benefit of the F function is that the sample size m can be varied, with larger values of m giving a smoother curve. Comparison of the G and F functions for the same point pattern may be informative since the two functions bring out different characteristics of the point pattern (O'Sullivan and Unwin, 2002).

The most widely used measure of spatial dependence in point patterns is the K function, which is the subject of the following section.

7.4.2 K function

Whereas the G and F functions are based on nearest neighbours, the K function is based on distances between all events in the region of study. The K function for distance d can be given by:

$$\hat{K}(d) = \frac{|A|}{n^2} \sum_{i=1}^{n} \#(C(\mathbf{x}_i, d)) \tag{7.12}$$

As defined previously, $\#(C(\mathbf{x}_i, d))$ indicates the number (#) of events in the circle $C(\mathbf{x}_i, d)$, which has as its centre the location $\mathbf{x}_i$ and radius d, and $|A|$ indicates the area of the study region. Recall that the hat (^) indicates that it is an estimate.

The K function can be computed by following several steps:

1. Go to an event and count the number of other events within a set radius of the event.

2. Do the same for all other events, adding the number of points within that set radius to the number of points in that radius for all events visited.

3. Once all events have been visited, the total counts within the distance band, d, is scaled (multiplied) by $\frac{|A|}{n^2}$.

4. Increase the radius a fixed amount and go back to step 1, repeating the process to the maximum desired radius.

As an example, for PP1 (Figure 7.1) there were five events within 5 units of the first event visited, so the unscaled version of $K(5)=5$. There were four events within 5 units of the second event visited, so the unscaled version of $K(5)$ becomes 9 and so on until all events have been visited.

For PP2 (Figure 7.2), the total number of events within 5 units of all events was zero (i.e. there were no other events within 5 units of any event) and for PP1 it was 122. As noted above, for each distance, the count is multiplied by $\frac{|A|}{n^2}$. The area of the study region is 6665.744 square units and there are 55 events in both PP1 and PP2. For PP1 $\hat{K}(5)$ is therefore obtained from:

$$\frac{6665.755}{55^2} \times 122 = 268.833$$

and for PP2, $\hat{K}(5)$ is:

$$\frac{6665.755}{55^2} \times 0 = 0$$

The expected value of K is given by:

$$E[K(d)] = \frac{\lambda \pi d^2}{\lambda} = \pi d^2 \tag{7.13}$$

where λ is the average intensity of the point pattern (but is not needed to calculate the expected value of K). Equation 7.13 gives the expected value for a point pattern that is said to be the outcome of a CSR process (see Section 7.3.1). A value of $K(d)$ greater than the expected value (i.e. πd^2) suggests a clustered point pattern, while a value of $K(d)$ less than the expected value suggests a regular point pattern. In practice, the L function, a transformed version of the K function, is often computed. The L function for distance d is given by:

$$\hat{L}(d) = \sqrt{\frac{\hat{K}(d)}{\pi}} - d \tag{7.14}$$

Values of the L function of less than 0 indicate regularity, while values greater than 0 suggest clustering. The K function is illustrated in Figure 7.13 using the point patterns illustrated in Figures 7.1 (PP1) and 7.2 (PP2), with the expected values ($E(K)$) also plotted. In this example, the smallest distance band was 2.5 units and this was increased in steps of 2.5 units to a maximum of 25 units. The L function is illustrated in Figure 7.14.

Note that the values of K for PP1 are clearly larger than those for PP2 and the expected value for smaller (≤ 17.5 units) distances, while the expected values are larger than those for PP1 for distances of 20 and greater. In contrast, values for PP2 are clearly smaller than the values for both PP1 and the expected values up to a distance band of 20 units. In words, PP1, defined as a 'clustered' point pattern, is indeed clustered at short distances, while PP2, defined as a 'regular' point pattern, is dispersed at small

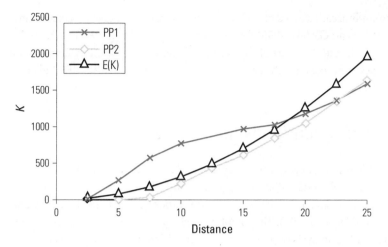

Figure 7.13 *K* function and expected values (*E*(*K*)) for the point patterns in Figures 7.1 (PP1) and 7.2 (PP2).

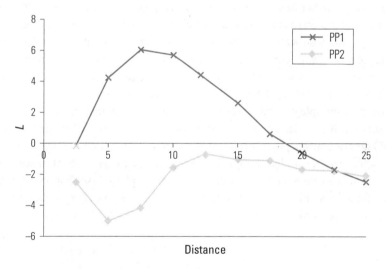

Figure 7.14 *L* function for the point patterns in Figures 7.1 (PP1) and 7.2 (PP2).

distances. In an application that makes use of the *K* function, O'Brien *et al.* (2000) demonstrate how different forms of cancers in humans and dogs tend to cluster over particular distance ranges. Such tools are a powerful means of assessing the structure of point patterns and they may provide evidence to support further analyses (e.g. if disease events cluster at particular scales we are likely to be interested in interpreting this finding).

An important issue in point pattern analyses concerns edge effects. Particularly for larger distance bands, parts of the circles drawn around points may often fall outside of the study area. To avoid biased results, it is necessary to take this factor into account (see Bailey and Gatrell (1995) for a discussion about this issue). For example, if KE is being conducted, and part of the kernel falls outside of the study area, then it will be necessary to account for the area of the kernel that is not within the study area.

7.5 Applications and other issues

Many different properties can be represented as point patterns. These include disease events (e.g. Hill *et al.*, 2000), trees (e.g. Li and Zhang, 2007), and concentrations of minerals in rock (Lloyd, 2006). Other applications are detailed by Bailey and Gatrell (1995) and Diggle (2003). This chapter provides only a brief outline of some key techniques for the analysis of point patterns. Consideration of the population at risk (e.g. accounting for the fact that disease rates tend to be higher in urban areas as there is a greater density of people in such areas than elsewhere), edge effects, and testing of point patterns (e.g. assessing how far a point pattern is clustered) are mentioned only quite briefly. Further issues that are not covered include the extension of the analysis of spatial point patterns to include a time element, analysis of marked point patterns (i.e. points with values attached), and techniques for the identification of clusters (i.e. specific locations with clusters as opposed to clustering of point patterns in general). Another issue that is not explored is the use of Monte Carlo randomization procedures in assessing point patterns relative to complete spatial randomness. With such approaches, multiple point patterns that are CSR are generated and these can be used to assess significant departures of real point patterns from CSR. Bailey and Gatrell (1995) describe such a procedure with respect to the L function. For the L function, it is common practice to derive approximate upper and lower 5% confidence intervals from simulated values, and these values can then be plotted (see Fotheringham *et al.* (2000) for an example). The next section presents a case study that makes use of data provided on the book website.

7.6 Case study

The following case study is based on locations in England that are represented on the Gough Map of Great Britain (dating to *c.* 1360). The data were captured as part of a research project funded by the British Academy[1] (see Lloyd and Lilley, 2009). The aim of the project was to develop a digital representation of the Gough Map. The analysis entailed using kernel estimation to explore local variation in event (i.e. place) intensity. Such an analysis is useful in that the results can be compared to those obtained using other data sources (e.g. contemporary records of taxation, etc.) to help

1 http://www.qub.ac.uk/urban_mapping/gough_map/

assess how representative the selection of places on the Gough Map is. For example, is the intensity of events (i.e. places) different for different sources and does this tell us anything about the purpose of these sources or where they were compiled? The intensity map was generated using ArcGIS™ Spatial Analyst. The kernel radius was set to 25 km and the cell size was 5 km. ArcGIS uses a kernel based on that defined by Silverman (1986, Equation 4.5). This is the same form as the kernel defined in Equation 7.7. The output gave intensities in square metres and these values were multiplied by one million to give intensities in square kilometres (as $1000 \, \text{m} \times 1000 \, \text{m} = 1,000,000 \, \text{m}^2$). The locations are shown in Figure 7.15 and the map of intensities is given in Figure 7.16.

The greater intensity in some regions (e.g. the south-east and mid- to north-east of England) is partly a function of varying numbers of places in England during the

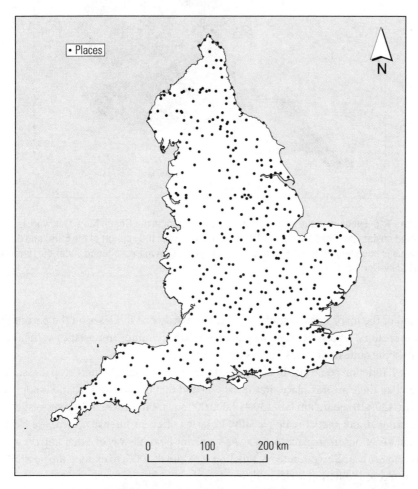

Figure 7.15 Places in England shown on the Gough Map. This work is based on data provided through EDINA UKBORDERS with the support of the ESRC and JISC and uses boundary material which is copyright of the Crown (reproduced under the terms of the Click-Use Licence).

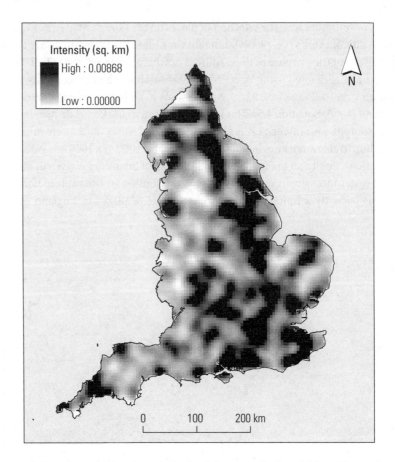

Figure 7.16 Event intensity for places in England shown on the Gough Map. This work is based on data provided through **EDINA UKBORDERS** with the support of the ESRC and JISC and uses boundary material which is copyright of the Crown (reproduced under the terms of the Click-Use Licence).

period of the map's construction, but it is also partly due to biases on the part of the map's creators, who selected some places of relatively minor importance at the time, while some quite high status centres do not appear on the map.

The *K* function could also be computed to assess clustering at different spatial scales, given that there are 461 places in the data set and that the area of England is approximately 130,410 square km (i.e. 130,410,000,000 square m), there is little suggestion of clustering at any spatial scale *globally*. In terms of event intensity, there are clearly spatial variations from place to place. As such, one possible way of extending the analysis is to use a *locally* computed *K* function (see Lloyd, 2006) to explore the possibility of clustering at different scales in different regions.

The synthetic data shown in Figures 7.1 and 7.2 are provided on the book website as is a guide to how the *L* function (see Figure 7.14) can be computed using these data with freely available software.

Summary

This chapter has discussed visual examination of spatial point patterns, summary measures, measures of event intensity, measures based on distances between events (event–event distances), and point–event distances, as well as some other issues. The focus has been on the description of point patterns in terms of the degree to which they are clustered or dispersed/regular. The techniques detailed enable extraction of information on various aspects of point patterns, including spatial variation in event intensity and spatial dependence in point patterns.

Further reading

Other introductions to point pattern analysis are provided by Bailey and Gatrell (1995), Fotheringham *et al.* (2000), O'Sullivan and Unwin (2002), Diggle (2003), Waller and Gotway (2004), and Lloyd (2006). Each of these accounts provides information on the methods detailed in this chapter as well as information on other issues, including analysis of space–time point patterns and testing of point patterns. A good summary of some approaches to the identification of clusters is given by Waller and Gotway (2004).

➡ The next chapter also deals with analysis of spatial patterning, but with data that have attributes (e.g. values attached to them) rather than the simple point events that were the focus in this chapter.

8

Exploring spatial patterning in data values

8.1 Introduction

This chapter introduces a variety of methods for the analysis of spatial variation in single and multiple variables. Methods are introduced that allow for the exploration in changes in values from place to place or in the way in which variables are related. In the first case, an example problem might be to ascertain if zones tend to be more similar to their neighbours in some parts of the study area than in others (e.g. do neighbourhoods in some areas have similar characteristics while neighbourhoods in other areas are quite different). In the second case, we may want to address questions such as 'How does the relationship between altitude and precipitation vary from place to place?' (e.g. does altitude seem to have an effect on precipitation amount in some areas but not others). The initial focus of the chapter is on the analysis of spatial struc- ture (spatial autocorrelation, i.e. the degree to which values at one location are similar to values at neighbouring locations). Next, the chapter moves on to computation of local statistics. The initial concern is with univariate measures; next, regression and correlation procedures are outlined that enable exploration of spatial relations between multiple variables. Finally, some other approaches are mentioned briefly before the chapter is summarized.

8.2 Spatial autocorrelation

Section 4.8 introduced the concepts of spatial autocorrelation and spatial dependence. Recall that spatial autocorrelation refers to the nature of correlation between neigh- bouring values while spatial dependence suggests the case where neighbouring values are similar (positive spatial autocorrelation specifically). A key tool for the analysis of spatial autocorrelation, the Moran's I coefficient, was introduced in Section 4.8. A locally derived version of Moran's I is detailed in Section 8.4.1. There are various other spatial autocorrelation measures that are applied widely. These include Geary's C, amongst others (see Bailey and Gatrell, 1995). Measures like Moran's I and Geary's C are conventionally used to explore spatial autocorrelation with neighbours of observations—that is, they enable assessment of the degree to which values tend to be similar to neighbouring values. It is also possible to explore how spatial autocorrelation varies with the distance separating observations (e.g. we can use a geographical weighting approach). One very useful tool for the exploration of spatial autocorrelation is the Moran scatter plot. The Moran scatter plot relates individual values to weighted averages of neighbouring values and the slope of a regression line fitted to the points in the scatter plot gives global Moran's I. An application of the Moran scatter plot is detailed in the case study in Section 8.7.1.

8.3 Local statistics

Section 4.6 introduced the idea of moving windows, whereby any statistic could be computed locally using a subset of the data. Section 4.7 extended this idea through the concept of geographical weights and inverse distance weighted prediction was illustrated. Another geographically weighted approach was demonstrated in Section 7.3.2, which introduced kernel estimation. In the following section, the idea of locally derived statistics is explored further, with a particular focus on local measures of spatial autocorrelation. Section 8.5 outlines some approaches to exploring local variations in the relationships between different variables.

8.4 Local univariate measures

Standard univariate statistical measures are often computed within a moving window, as demonstrated previously in Section 4.6. Such measures allow exploration of the degree and nature of variation in summary statistics across the region of interest. For example, a local version of the standard deviation enables assessment of the degree of variability in the property of interest from place to place. Knowledge of such variation is often crucial in interpreting spatial data. Section 10.4 explores this issue further with a focus on raster grid data.

Geographical weighting schemes are widely used in the estimation of local statistics. The quartic kernel, illustrated in the previous chapter, includes bandwidth, τ, which determines the degree of weighting by distance. For a small bandwidth, locations close to the centre of the window will receive most of the weight. In contrast, for a large bandwidth, more distant locations will also receive quite large weights. A large bandwidth therefore corresponds to a wide 'hump' and a small bandwidth corresponds to a narrow 'hump'. Another widely used weighting scheme is the Gaussian weighting scheme. As in previous chapters, the weight for location i can be given by w_{ij}, indicating the weight of sample i with respect to location j. The Gaussian weighting scheme is given by (Fotheringham et al., 2002):

$$w_{ij} = \exp\left[-0.5\left(\frac{d_{ij}}{\tau}\right)^2\right] \tag{8.1}$$

which indicates that the weight for location i with respect to location j is obtained by multiplying -0.5 by the square of the distance d between locations i and j (i.e. d_{ij}) divided by the bandwidth τ and then obtaining the exponential value of the product (this can easily be computed in a spreadsheet package, where 'exp' is the standard abbreviation for the exponential function). As an example, for a bandwidth of 10 and a distance of 15:

$$\exp\left[-0.5\left[\frac{15}{10}\right]^2\right] = \exp[-0.5(1.5)^2] = \exp[-0.5 \times 2.25] = \exp[-1.125] = 0.3247$$

where $\exp[-1.125]$ can be obtained with $2.718281828^{-1.125}$ ($= 1/2.718281828^{1.125}$) and 2.718281828 is the approximate base of the natural logarithm (see Wilson and Kirkby (1980) for more details). Appendix B shows one way to compute the exponential function.

The Gaussian weighting scheme, for bandwidths of 5, 10, and 15 units, is illustrated in Figure 8.1. Note that, unlike the case for the quartic kernel, the bandwidth for the Gaussian weighting scheme does not extend to the outer edge of the kernel, but of course the bandwidth still determines the kernel size.

Any standard statistic can be geographically weighted (see Fotheringham et al. (2002) for more information). As an example of this geographical weighting scheme in practice, obtaining the locally weighted mean using the Gaussian weighting scheme is illustrated below. Following Section 4.7, the locally weighted mean is given by:

$$\bar{z}_i = \frac{\sum_{j=1}^{n} z_j w_{ij}}{\sum_{j=1}^{n} w_{ij}} \tag{8.2}$$

Table 8.1 and Figure 8.2 detail the locations of a set of observations which are the same as those used in Section 4.7. However, in this case, the first observation is treated as known. The weights, obtained using the Gaussian weighting scheme detailed above (with bandwidths of 5, 10, and 15 units), are given for each distance. The weights for each location are then multiplied by the value at that location. As an example, following Equation 8.1 for a bandwidth of 10 the products of the multiplications are summed, giving a value of 136.733. The weight values are also summed, giving a value of 6.643. The weighted mean is then obtained by $136.733/6.643 = 20.582$. Given the 20 observations, the unweighted mean is 19.050.

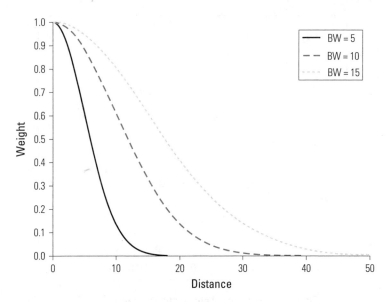

Figure 8.1 Gaussian weighting scheme: bandwidths of 5, 10, and 15 units.

Table 8.1 Observations (j), distance from observation 1 (d_{ij}), weights (w_{ij}) and weights multiplied by values ($z_j w_{ij}$)

			$\tau=5$		$\tau=10$		$\tau=15$	
j	d_{ij}	z_j	w_{ij}	$z_j w_{ij}$	w_{ij}	$z_j w_{ij}$	w_{ij}	$z_j w_{ij}$
1	0.000	9	1.000	9.000	1.000	9.000	1.000	9.000
2	4.404	14	0.679	9.499	0.908	12.706	0.958	13.409
3	9.699	43	0.152	6.553	0.625	26.867	0.811	34.889
4	10.408	12	0.115	1.375	0.582	6.981	0.786	9.433
5	10.871	34	0.094	3.198	0.554	18.829	0.769	26.147
6	12.958	26	0.035	0.905	0.432	11.230	0.689	17.903
7	13.959	24	0.020	0.487	0.377	9.059	0.649	15.565
8	14.066	33	0.019	0.631	0.372	12.271	0.644	21.260
9	15.506	34	0.008	0.277	0.301	10.219	0.586	19.927
10	17.256	10	0.003	0.026	0.226	2.256	0.516	5.160
11	17.606	8	0.002	0.016	0.212	1.698	0.502	4.017
12	18.018	13	0.002	0.020	0.197	2.564	0.486	6.319
13	18.025	11	0.002	0.017	0.197	2.167	0.486	5.344
14	18.285	24	0.001	0.030	0.188	4.510	0.476	11.416
15	19.253	9	0.001	0.005	0.157	1.410	0.439	3.949
16	21.335	15	0.000	0.002	0.103	1.541	0.364	5.455
17	23.845	14	0.000	0.000	0.058	0.816	0.283	3.957
18	23.988	34	0.000	0.000	0.056	1.914	0.278	9.465
19	24.464	3	0.000	0.000	0.050	0.150	0.264	0.793
20	24.522	11	0.000	0.000	0.049	0.544	0.263	2.891
Sum		381	2.132	32.042	6.643	136.733	11.248	226.300
Mean		19.050		15.032		20.582		20.119

Weights are obtained using the Gaussian weighting scheme with a bandwidth (τ) of 5, 10, and 15 units.

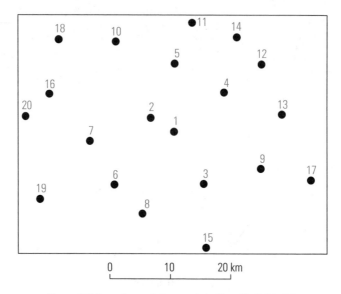

Figure 8.2 Locations of observations listed in Table 8.1.

With reference to Table 8.1, note how the weights for large distances are proportionately smaller when the bandwidth is smaller—that is, a small bandwidth gives most influence to close-by observations whereas with a large bandwidth more distant observations have proportionately larger weights. The weighted means (or other statistics) could be calculated anywhere—at the location of an observation or anywhere else (as for the inverse distance weighting example in Section 4.7). Fotheringham *et al.* (2002) discuss a range of geographically weighted statistics.

The spatial scale of a process can be explored using geographical weights. For example, by assessing the results obtained using a variety of different kernel bandwidths, it is possible to explore how much these results vary and, therefore, learn something about dominant scales of spatial variation. If the geographically weighted mean average changes a great deal as the bandwidth is increased for small bandwidths, but then stabilizes as the bandwidth is increased to some critical value, then we can say (with some caveats) that most variation in the property of concern is at some scale finer than that represented by the bandwidth distance at which results stabilize.

The following section shows how spatial autocorrelation measures can be computed locally.

8.4.1 Local spatial autocorrelation

In most published applications, spatial autocorrelation is measured over the entire study area, as was detailed in Section 4.8. However, such an approach masks any spatial variation in the spatial structure of the variable of interest. For this reason, various local measures of spatial autocorrelation have been developed. One of the most widely used is a local variant of Moran's I presented by Anselin (1995). It is given by:

$$I_i = z_i \sum_{j=1}^{n} w_{ij} z_j, \; j \neq i \tag{8.3}$$

where z_j are differences of variable y from its global mean ($y_i - \bar{y}$). In cases where zones are used (as opposed to points) the weights, w_{ij}, are often set to 1 for immediate neighbours of a zone and 0 for all other zones. Local I often appears in modified form:

$$I_i = \left[\frac{z_i}{s^2}\right] \sum_{j=1}^{n} w_{ij} z_j, \; j \neq i \tag{8.4}$$

where s^2 is the sample variance (the square of Equation 3.3). Note that local I values sum up to global Moran's I. Anselin (1995) describes an approach to testing for significant local autocorrelation based on random relocation of the data values, the objective being to assess if the observed configuration of values is significant. The GeoDa software offers the capacity to test the significance of local I using randomization[1] and to map significant clusters. Clusters are identified using the Moran scatter plot (Anselin, 1996).

Local I is demonstrated following Equation 8.4 using the following grid:

45 44 44

43 42 39

38 32 34

Local I is computed for the central cell and rook's case weighting is used—only the cells in the same row or column are used. The mean of values in the entire data set is needed first. Here there are only nine values and their mean is 40.111 and the sample variance is 21.861. Table 8.2 shows the original values (y_j), their deviations from the mean (z_j), the weights (w_{ij}), and weights multiplied by the deviations from the mean ($w_{ij}z_j$).

In this case the weights are row standardized, i.e. they sum to 1 (there are four values of 0.25 and these are for the four cells which share an edge with the central cell, which has a value of 42). z_i is 1.889, the sum of the weights multiplied by the deviations from the mean ($w_{ij}z_j$) is −0.611, as shown in Table 8.2. I_i is then given by ($1.889/21.861$)×−0.611 = −0.053. In this case, I_i has a negative value, indicating negative spatial autocorrelation, i.e. neighbouring values tend to be dissimilar.

Figure 8.3 gives a map of the log of the number of persons per hectare in Northern Ireland in 2001; the values were logged as the raw population densities had a positively skewed distribution and the transformed values have almost zero skew. Figure 8.4 gives an example of the application of I_i for measuring spatial autocorrelation (using queen's case contiguity; the use of contiguity schemes with non-gridded data was outlined in Section 4.8) in logged population density given the data shown in

---

1 see https://www.geoda.uiuc.edu/support/help/glossary.html

Table 8.2 Values, differences from the mean, and weights

y_i	z_i	w_{ij}	$w_{ij}z_j$
45	4.889	0.000	0.000
43	2.889	0.250	0.722
38	−2.111	0.000	0.000
44	3.889	0.250	0.972
42	1.889	0.000	0.000
32	−8.111	0.250	−2.028
44	3.889	0.000	0.000
39	−1.111	0.250	−0.278
34	−6.111	0.000	0.000
Sum		1.000	−0.611

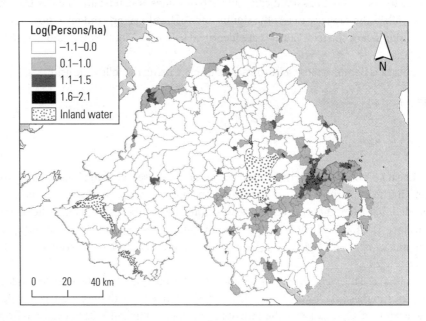

Figure 8.3 Log of persons per hectare in Northern Ireland in 2001. Northern Ireland Census of Population data—© Crown Copyright. Reproduced under the terms of the Click-Use Licence.

Figure 8.3. Global I was 0.665. In the case of I_i, there are large positive values in rural areas (which have larger zones, since zone size is a function of the population density). In urban areas like Belfast (in the east), the contrast between central areas with high population densities and suburban and rural areas with lower population densities are apparent. Areas with contrasting values are marked by small positive or negative values of I_i. Note that the results are a function of the particular form of zones used (see Section 4.9 for a relevant discussion). There are many published applications of local I (e.g. Anselin, 1995; Lloyd, 2006).

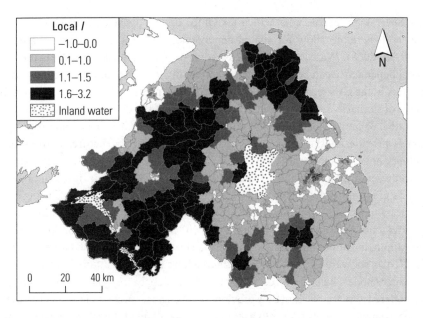

Figure 8.4 Local *I* for log of persons per hectare in Northern Ireland in 2001 using queen's case contiguity. Northern Ireland Census of Population data—© Crown Copyright. Reproduced under the terms of the Click-Use Licence.

The focus of the chapter now moves onto analysis of spatial patterning in the relationships between multiple variables.

8.5 Regression and correlation

The subject of regression and correlation was introduced in Section 3.3. Some regression-based analyses of spatially referenced variables map the residuals (in the two-variable case, this means the difference between the value indicated by the line of best fit and the observed value) from the regression (see the example in Figure 3.5). It is straightforward to take this a step further and explore not just how accurate fitted values are from place to place but to consider how the relationships between variables differ spatially. As with univariate statistics, multivariate approaches (such as correlation and regression) can be conducted in a moving window. The next section introduces some approaches to regression in the analysis of spatial data. Next, the focus is on regression conducted in a moving window. Following this, a geographically weighted approach is detailed and this account makes use of matrix algebra to obtain regression coefficients.

8.5.1 Spatial regression

An assumption of standard ordinary least squares regression is independence of observations. As discussed in Section 3.5, this assumption rarely holds true for spatial data. Several approaches exist which take into account the spatial structure in variables and therefore allow for spatial dependence (e.g. Rogerson, 2006; Ward and Gleditsch, 2008). With generalized least squares (GLS) regression, information on spatial dependence can be utilized when estimating the regression coefficients. More specifically, the GLS regression coefficients can be estimated given information on the degree of similarity between variables as a function of the distance by which they are separated. Bailey and Gatrell (1995) provide an account of GLS. Spatial autoregressive models provide another means of accounting for spatial structure. Lloyd (2006) outlines the simultaneous autoregressive model, which includes an interaction parameter representing interactions between neighbouring observations. If the interaction parameter is unknown (the usual case), then such models cannot be fitted using ordinary least squares and specialist software is required. The simultaneous estimation of the interaction parameter and the regression (β) coefficients can be conducted using a maximum likelihood procedure, as described by Schabenberger and Gotway (2005). The GeoDa software of Anselin *et al.* (2006) allows for spatial autoregressive modelling. Such models help to overcome the problem of analysing relations between spatially referenced variables, but they provide only a single set of coefficients. Increasingly, in GIS contexts, studies take into account the local context. Local approaches entail estimating regression coefficients using either local data subsets or a geographical weighting scheme. Two local regression approaches are outlined next.

8.5.2 Moving window regression

Section 8.4.1 introduced the idea that spatial autocorrelation of variables is often observed to vary spatially. Local measures which take these variations into account may therefore be worthwhile. Similarly, relationships between variables may differ markedly across an area. As an example, many studies have shown that altitude and precipitation amount are related in some regions. However, while the two variables may be strongly related in some areas, a global regression of altitude and precipitation amount may demonstrate only a weak relationship (see Lloyd (2005) for a relevant case study). Some kind of local regression procedure is therefore needed to enable exploration of some geographically variable relationships.

One straightforward approach to exploring how relationships vary spatially is simply to conduct a standard regression in a moving window. In other words, regression is carried out using only the data in the moving window and the end result is a set of maps of regression coefficients. Moving window regression (MWR) has been used in various studies (see Lloyd (2005, 2006) and Lloyd and Shuttleworth (2005) for examples). MWR is identical to the regression procedure detailed in Section 3.3, the only difference being that regression is conducted for data subsets in a moving window rather than for all data simultaneously. Building on the geographical weighting principles outlined previously (see Section 4.7), this approach can be extended such that the influence of observations in the regression is decreased as distance from the centre of the moving window increases. Such an approach is the subject of the following section.

8.5.3 Geographically weighted regression

This book has outlined various geographically weighted statistics and argued that such approaches are intuitively sensible since we expect places close together to be more alike than places a greater distance apart. Geographically weighted regression (GWR) extends the same principle to regression analysis. GWR has become a core tool in many analyses of spatial data. The approach is described in detail by Fotheringham *et al.* (2002) but an account of some key principles is given here. Essentially, the key steps in a GWR analysis are as follows:

1. Go to a location (zone or point).

2. Conduct regression using all data (or some subset) but give greater weight (influence) to locations that are close to the location of interest—a geographical weighting scheme is used.

3. Move to the next location and go back to stage 2 until all locations have been visited.

The output is a set of regression coefficients (e.g. for bivariate regression (with one independent and one dependent variable), the intercept, and slope) at each location. GWR coefficients are obtained using:

$$\beta(\mathbf{x}_i) = (\mathbf{Y}^T \mathbf{W}(\mathbf{x}_i) \mathbf{Y})^{-1} \mathbf{Y}^T \mathbf{W}(\mathbf{x}_i) \mathbf{z} \tag{8.5}$$

This is the same as for ordinary unweighted regression (Equation 3.9), except that the regression coefficients are computed for each location $\mathbf{x}_i$ and there are geographical weights given by $\mathbf{W}(\mathbf{x}_i)$. If all weights were equal to 1 then this would correspond to standard unweighted regression. The weights matrix is given by:

$$\mathbf{W}(\mathbf{x}_i) = \begin{bmatrix} w_{i1} & 0 & 0 & 0 \\ 0 & w_{i2} & 0 & 0 \\ 0 & 0 & \ddots & 0 \\ 0 & 0 & 0 & w_{in} \end{bmatrix}$$

w_{i1} is the weight given the distance between the location i and observation 1. The diagonal dots ($\ddots$) indicate that the matrix can be expanded—that is, if n (the number of observations) is 5 then the matrix will have 5×5 entries, with non-zero values only in the diagonal of the matrix.

One weighting scheme used widely in GWR contexts is the Gaussian weighting scheme, detailed in Equation 8.1. Note that MWR is a special case of GWR where the weights for the n nearest neighbours are set to 1 and all other weights are 0.

GWR is illustrated using the data listed in Table 8.3 and mapped in Figure 8.5. Table 8.3 gives the coordinates of the observations, variable 1 (independent) and variable 2 (dependent) values, distance from the first observation, and geographical weights ('Geog. wt.'; using the Gaussian weighting function) for a bandwidth of 10 units. The computation

Table 8.3 Coordinates of observations, variable 1 and 2 values, distance from the first observation, and geographical weights

No.	x coordinate	y coordinate	Variable 1 (y)	Variable 2 (z)	Distance (d_{ij})	Geog. wt. (w_{ij})
1	25.00	45.00	12	6	0.00	1.0000
2	25.51	44.14	34	52	1.00	0.9950
3	21.87	48.90	32	41	5.00	0.8825
4	27.60	52.57	12	25	8.00	0.7261
5	16.69	31.33	11	22	16.00	0.2780
6	42.52	35.35	14	9	20.00	0.0889
7	9.20	65.65	56	43	26.00	0.0340
8	29.23	76.72	75	67	32.00	0.0060
9	61.37	66.01	43	32	42.00	0.0002

Bandwidth = 10 units.

of geographical weights using Equation 8.1 was illustrated in Section 8.4. Note that the number of observations is small and this example is used purely for ease of illustration.

In this example, the weight matrix is therefore:

$$\mathbf{W}(\mathbf{x}_i) = \begin{bmatrix} 1.0000 & 0 & 0 & 0 & 0 & 0 & 0 & 0 & 0 \\ 0 & 0.9950 & 0 & 0 & 0 & 0 & 0 & 0 & 0 \\ 0 & 0 & 0.8825 & 0 & 0 & 0 & 0 & 0 & 0 \\ 0 & 0 & 0 & 0.7261 & 0 & 0 & 0 & 0 & 0 \\ 0 & 0 & 0 & 0 & 0.2780 & 0 & 0 & 0 & 0 \\ 0 & 0 & 0 & 0 & 0 & 0.0889 & 0 & 0 & 0 \\ 0 & 0 & 0 & 0 & 0 & 0 & 0.0340 & 0 & 0 \\ 0 & 0 & 0 & 0 & 0 & 0 & 0 & 0.0060 & 0 \\ 0 & 0 & 0 & 0 & 0 & 0 & 0 & 0 & 0.0002 \end{bmatrix}$$

The regression coefficients are obtained as for the standard regression approach detailed above except that $\mathbf{Y}^T$ is multiplied by $\mathbf{W}(x_i)$. Following this procedure (by referring to the example for global regression in Section 3.3 and Appendix E it should be possible to work out what is going on):

$$\mathbf{Y}^T\mathbf{W}(\mathbf{x}_i) = \begin{bmatrix} 1 & 0.9950 & 0.8825 & 0.7261 & 0.2780 & 0.0889 & 0.0340 & 0.0060 & 0.0001 \\ 12 & 33.8300 & 28.2400 & 8.7132 & 3.0580 & 1.2446 & 1.9040 & 0.4500 & 0.0086 \end{bmatrix}$$

$$\mathbf{Y}^T\mathbf{W}(\mathbf{x}_i)\mathbf{Y} = \begin{bmatrix} 4.0107 & 89.4484 \\ 89.4484 & 2494.2646 \end{bmatrix}$$

$$(\mathbf{Y}^T\mathbf{W}(\mathbf{x}_i)\mathbf{Y})^{-1} = \begin{bmatrix} 1.2454 & -0.0447 \\ -0.0447 & 0.0020 \end{bmatrix}$$

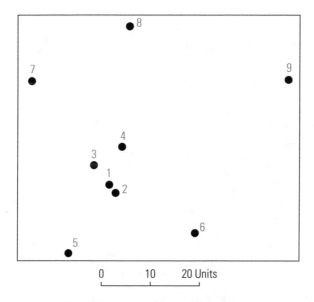

Figure 8.5 Locations of the data in Table 8.3.

$$\mathbf{Y}^T\mathbf{W}(\mathbf{x}_i)\mathbf{z} = \begin{bmatrix} 120.8615 \\ 3397.6046 \end{bmatrix}$$

$$\beta(\mathbf{x}_i) = \mathbf{Y}^T\mathbf{W}(\mathbf{x}_i)\mathbf{Y})^{-1}\mathbf{Y}^T\mathbf{W}(\mathbf{x}_i)\mathbf{z} = \begin{bmatrix} -1.223 \\ 1.406 \end{bmatrix}$$

The intercept, $\beta_0(\mathbf{x}_i)$, is therefore -1.223 and the slope, $\beta_1(\mathbf{x}_i)$, is 1.406. As an example, for the first case in Table 8.3, the fitted value is $-1.223 + 1.406 \times 12 = 15.649$.

Figure 8.6 shows the ordinary (unweighted) and geographically weighted regression lines fitted with the geographical weights indicated. Note the effect of geographical weighting on, in effect, pulling the line towards the points with larger weights. You can match the geographical weight and the values of the two variables to the entries in Table 8.3.

The goodness of fit of the GWR locally can be assessed using the geographically weighted coefficient of determination (r^2) (Fotheringham et $al.$, 2002). The geographically weighted r^2 for location i is given by:

$$r_i^2 = \frac{\mathrm{TSS}_i - \mathrm{RSS}_i}{\mathrm{TSS}_i} \qquad (8.6)$$

where TSS_i is the geographically weighted total sum of squares given by:

$$\mathrm{TSS}_i = \sum_{j=1}^{n} w_{ij}(z_j - \bar{z})^2$$

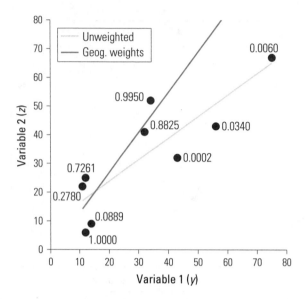

Figure 8.6 Regression using the data in Table 8.3: unweighted and geographically (geog.) weighted, with geographical weights indicated.

This is the sum of the weights multiplied by the squared difference between each (dependent variable) value and its mean. RSS_i is the geographically weighted residual sum of squares given by:

$$RSS_i = \sum_{j=1}^{n} w_{ij}(z_j - \hat{z}_j)^2$$

This is the sum of the weights multiplied by the squared residual (the difference between each value and the value given the GWR model).

The calculations following Equation 8.6 are presented in Table 8.4.

In this case, each term is as follows:

$$TSS_i = \sum_{j=1}^{n} w_{ij}(z_j - \bar{z})^2 = 1286.326$$

$$RSS_i = \sum_{j=1}^{n} w_{ij}(z_j - \hat{z}_j)^2 = 266.227$$

$$r_i^2 = \frac{TSS_i - RSS_i}{TSS_i} = \frac{1286.326 - 266.227}{1286.326} = 0.7930$$

The unweighted r^2 is 0.7413 and the geographically weighted r^2 is 0.7930 (or 0.7966 calculated using purpose-written software, the difference being due to rounding errors). In other words, the GWR model is a better fit than the unweighted model in this case and this suggests that taking into account distance from the location of

Table 8.4 Calculations for the geographically weighted coefficient of determination (data as for Table 8.3), with mean $(\bar{z})$ of 33

Obs. (j)	$w_{ij}(z_j - \bar{z})^2$	$\hat{z}_j$	$w_{ij}(z_j - \hat{z}_j)^2$
1	729.000	15.649	93.103
2	359.195	46.581	29.219
3	56.480	43.769	6.766
4	46.470	15.649	63.491
5	33.638	14.243	16.728
6	51.206	18.461	7.957
7	3.400	77.513	40.499
8	6.936	104.227	8.315
9	0.000	59.235	0.148
Sum	1286.326		266.227

Obs., observation.

interest is beneficial in this case. Again, it should be stressed that the number of observations is small for this example and this should be considered in any analysis.

The GWR software offers two approaches to assessing the significance of the GWR model. In essence, these approaches allow users to determine if any of the local parameter estimates are 'significantly non-stationary' (Fotheringham *et al.*, 2002, p. 213), where a non-stationary model is one which has parameters that vary geographically. In other words, the test enables assessment of the degree to which the GWR model is an improvement over a standard global model.

To illustrate GWR, data on elevation and precipitation amount in Northern Ireland for July 2006 (with data at 149 locations) are analysed. The data locations are shown in Figure 8.7 and an interpolated (see Section 9.2) map of precipitation amounts is given in Figure 9.7. Figure 8.8 is a scatter plot for all observations (it is a global regression) with a fitted regression line. This indicates that the expected precipitation amount at a location with an altitude of 0 m is 52.203 mm (i.e. the intercept) and that this increases on average by 0.1162 mm (the slope) with an increase in altitude of 1 m. While the coefficient of determination (r^2) indicates that the model explains some 42% of the variation, there is clearly much variation around the regression line and GWR allows for exploration of local variations in this relationship.

A GWR bandwidth can be selected using cross-validation. Cross-validation entails removal of an observation, regression conducted using the remaining observations, and prediction of the removed value—that is, the regression equation is used to predict the value of z (the dependent) given a value of y (the independent), as described in Section 3.3. The removed value is then added back and the observation removed at the next location (in whatever order the locations are visited) after which the procedure is repeated for all remaining observations. The difference between the observed and predicted values is then computed. The bandwidth that results in the smallest cross-validation error is retained. An additional method for bandwidth selection, which is employed by the GWR software of Fotheringham *et al.* (2002), is called the

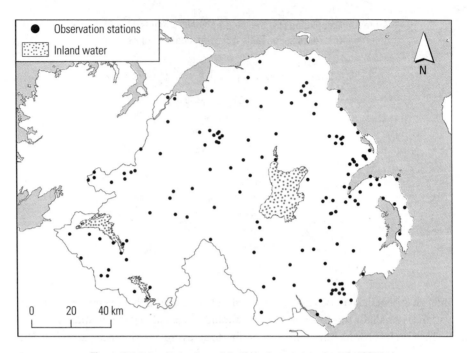

Figure 8.7 Locations of precipitation observations in July 2006.

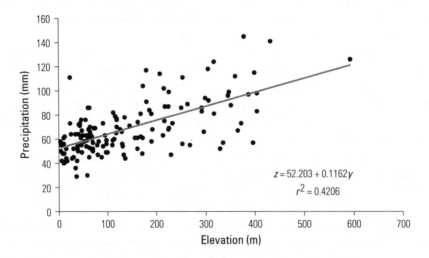

$$z = 52.203 + 0.1162y$$
$$r^2 = 0.4206$$

Figure 8.8 Elevation against July 2006 precipitation amount in Northern Ireland.

Akaike Information Criterion (AICc; Fotheringham *et al.*, 2002, Equation 4.21). The AICc allows for assessment of the number of degrees of freedom and the goodness of fit of the model and it can be used to compare the performance of a global regression model and GWR (Fotheringham *et al.*, 2002). The geographically weighted bandwidth of 9952.763 m was selected using the AICc in version 3.0 of the GWR software.

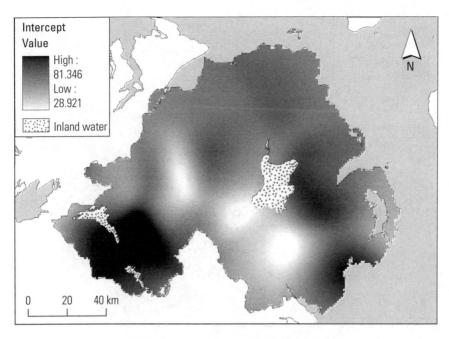

Figure 8.9 GWR intercept: elevation against July 2006 precipitation amount in Northern Ireland.

GWR can be conducted at any location, thus a GWR model can be fitted at a location where there is no observation. In the following example, GWR model parameters were obtained on a 1-km grid and this makes visualization of spatial variation in parameters easier than when parameter estimates are made at the locations of observations only. Figure 8.9 shows the mapped GWR intercept values, Figure 8.10 shows the slope coefficients, and Figure 8.11 shows the coefficients of determination (r^2; here computed outside of GWR 3.0, as that package does not return standard GW r^2 values). Note that the number of decimal places is kept at three for all of the following figures since the values for the slope and r^2 are small.

Clearly, there is much variation in the nature and the strength of the relationship between elevation and precipitation amount in Northern Ireland. Values of the GWR intercept (Figure 8.9) are largest (and larger than the global model intercept of 52.203 mm) in parts of the south west, parts of the mid east (south and west of Belfast), and the south east. Where the intercept is large and the slope has a negative or small positive value, this indicates that, in such areas, precipitation amounts are large (in proportion to the size of the intercept), irrespective of the elevation. Comparison of some areas in Figure 8.9 (GWR intercept) with Figure 8.10 (GWR slope) shows that some areas (e.g. in parts of the south west) fulfil these criteria. The GWR slope values (Figure 8.10) are largest in parts of the far west, the Ards Peninsula (south and west of Belfast), and some other areas, most notably to the south and west of Lough Neagh. Where the GWR r^2 (Figure 8.11) is also large, this suggests that elevations and precipitation amounts are strongly related (i.e. a large slope indicates a large increase in

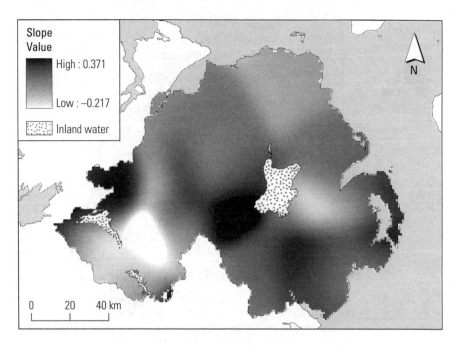

Figure 8.10 GWR slope parameter: elevation against July 2006 precipitation amount in Northern Ireland.

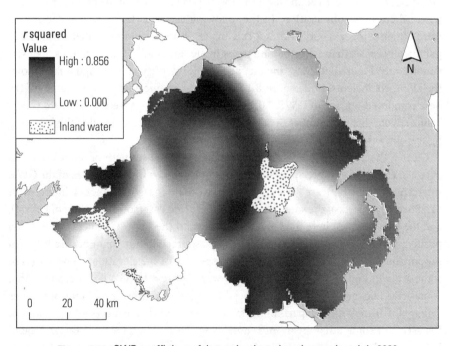

Figure 8.11 GWR coefficient of determination: elevation against July 2006 precipitation amount in Northern Ireland.

precipitation amount with an increase in elevation). Parts of the far west of the region have large slope and large r^2 values. The GWR slopes and r^2 values tend to be larger in areas with larger elevation values (see Figure 10.5 for a map of elevation in Northern Ireland) and this suggests that the elevation–precipitation relationship tends to be less strong (and therefore elevation is a less useful predictor) in areas with small elevation values. Trends in precipitation amounts are directional and are a function of many factors not considered here but, as the example demonstrates, GWR is a powerful means of exploring spatially variable relationships such as those between elevation and precipitation amount.

Relationships between many variables of interest in the physical and social sciences are a function of geography. These include the previous example of altitude and precipitation as well as other variables such as employment status and religion. Where such geographical variations are suspected, standard global regression analyses may be inadequate and an analysis based on the application of GWR may reveal a far richer picture than would be obtained through conventional regression analysis. GWR allows assessment of how far the nature of relationships (e.g. are variables related positively or negatively) vary spatially and how strongly variables are related in different regions. GWR has been used in many other contexts. Brunsdon *et al.* (2001) used GWR to explore the average elevation–precipitation relationship across Britain, while Lloyd and Shuttleworth (2005) used GWR to explore spatial variation in the relationship between commuting distance and a range of other variables in Northern Ireland. Some authors have commented on potential problems associated with GWR, particularly in terms of multicollinearity (i.e. strong correlations between independent variables). This may have an impact on the interpretation of the local regression coefficients (see Wheeler and Tiefelsdorf, 2005). In short, where there are multiple independent variables in the GWR analysis, and these independent variables are strongly related, the values and signs of coefficients for individual variables may be highly misleading. Methods for diagnosing collinearity and a solution to this problem are detailed by Wheeler (2007). Another issue which should be considered is the availability of enough observations with significantly non-negative weights for the purposes of GWR parameter estimation. One way around this problem is to use an adaptive bandwidth whereby the size of the bandwidth varies as a function of the density of observations in a given area (see Fotheringham *et al.*, 2002).

8.6 Other approaches

This chapter offers only a brief summary of some widely used approaches for the analysis of spatial structure in single and multiple variables. The selection is biased and many other approaches could have been included. For example, in terms of regression approaches, a body of models called multilevel models is used widely by geographers (and others) to explore relationships between variables at different spatial scales (see Fotheringham *et al.* (2002) and Lloyd (2006) for summaries of some other approaches).

The chapter introduces some key methods and concepts which will hopefully aid understanding of other approaches not considered here.

8.7 Case studies

This section comprises two case studies that make use of the same data set, which is provided on the book website. The data represent elevations in Switzerland and precipitation amounts for 8 May 1986; the number of observations is 467. The data are described by Dubois (2003) and the precipitation measurements are shown in Figure 8.12. The first case study makes use of Moran's I autocorrelation coefficient to explore spatial variation in precipitation amount. The second study uses GWR to explore spatial variation in the relationship between elevation and precipitation amount.

8.7.1 Spatial autocorrelation analysis

Moran's I was computed using geographical weights with the Gaussian weighting function defined in Equation 8.1. For a 10-km bandwidth, I was 0.722, indicating strong positive spatial autocorrelation in precipitation amounts. The data were further

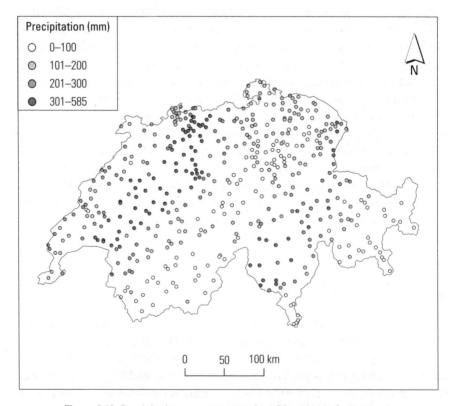

Figure 8.12 Precipitation measurements for 8 May 1986 in Switzerland.

interrogated through computing the Moran scatter plot, as described in Section 8.2, and this is shown in Figure 8.13. The plot shows the value at location i plotted against the weighted average of neighbouring values and the slope of the line fitted to the scatter plot gives global Moran's I. The plot allows assessment of outliers—those values that do not fit the general trend. For example, there are some points in the right-hand side of the upper right quadrant that fall far from the fitted line. Identifying these locations on a map could be informative, and this would be a sensible part of a fuller analysis. Moran's I could be computed using a variety of different bandwidths and changes in results with change in bandwidth help to suggest the scales over which the property is spatially autocorrelated.

An additional step that could be taken is to map the local I values, as in the case of Figure 8.4. Moran scatter plots and local I can both be computed using the GeoDa software (Anselin *et al.*, 2006).

8.7.2 Geographically weighted regression

For this analysis, the elevation data were provided as a digital elevation model (DEM) and the values at each precipitation observation location were extracted; the DEM is shown in Figure 10.17. A global regression of elevation and daily precipitation amount suggests that the two are only weakly related—the r^2 value was 0.0366 and the slope parameter coefficient was negative. Intuitively, we would expect elevation and precipitation to be positively related in at least some areas even for a period as short as a day (note that, for the example for Northern Ireland given in Section 8.5.3, the data were for monthly precipitation amounts). The GWR software (Fotheringham *et al.*, 2002)

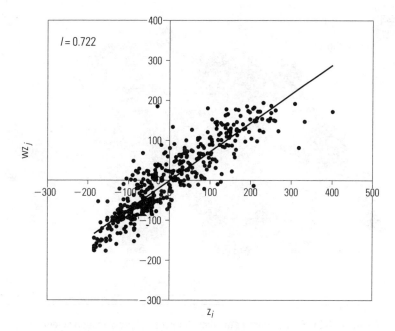

Figure 8.13 Moran scatter plot for precipitation measurements: 10-km Gaussian bandwidth.

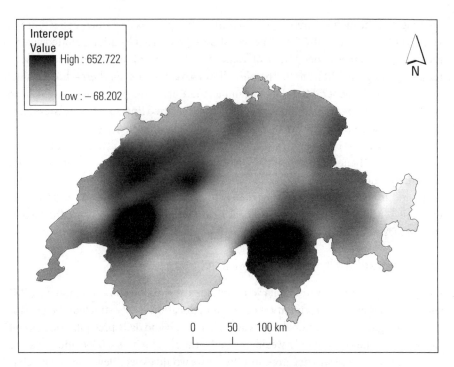

Figure 8.14 GWR intercept: elevation against precipitation amount for 8 May 1986 in Switzerland.

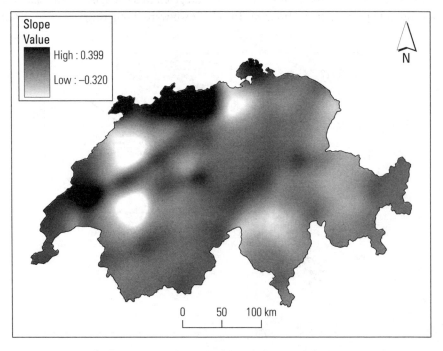

Figure 8.15 GWR slope parameter: elevation against precipitation amount for 8 May 1986 in Switzerland.

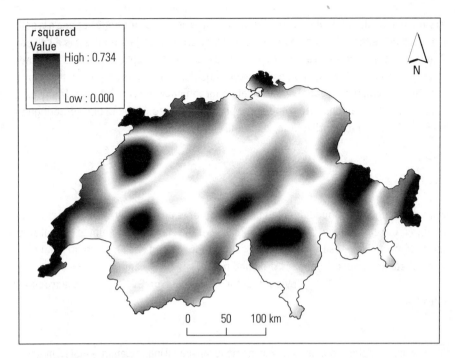

Figure 8.16 GWR coefficient of determination: elevation against precipitation amount for 8 May 1986 in Switzerland.

was used to fit models locally and assess how this relationship varies. The AICc was used to select a GWR kernel bandwidth of 11034.681 m. The GWR intercepts, slopes, and r^2 (as for the Northern Ireland data, the latter was computed outside of GWR 3.0) values are shown in Figures 8.14, 8.15, and 8.16, respectively.

Figures 8.14 and 8.15 provide information about the form of the relationship between elevation and precipitation amount, while Figure 8.16 suggests that the two are strongly related in some regions. Figure 8.15 indicates that the relationship between elevation and precipitation is negative (as for the global regression) or weakly positive in many areas. In some areas, however, there is a strong positive relationship, as indicated by large slope values (Figure 8.15) and large values of the coefficient of determination (r^2; Figure 8.16). The most obvious areas that fit within this category are in the north-west and the far west of Switzerland. The maps could be interpreted further with reference to the case study for Northern Ireland in Section 8.5.3.

Summary

This chapter has introduced approaches for deriving local statistics and for the analysis of spatial autocorrelation at different spatial scales and locally. The principal focus (in terms of space devoted to topics) has been on methods for the analysis of geographically varying relationships between variables. Knowledge of such approaches opens a wealth of opportunities for exploring spatial data. Exploratory spatial data analysis is a key part of any spatial analysis more generally, and assessing geographical variations in individual variables and in relationships between variables represents a significant improvement on conventional aspatial analyses of geographically referenced data. Many case studies exist that demonstrate some of the possibilities (see, for example, Fotheringham *et al.,* 2002; Lloyd and Shuttleworth, 2005; and Lloyd, 2006).

Further reading

More depth on the methods detailed in this chapter is provided by Fotheringham *et al.* (2000), Waller and Gotway (2004), and Lloyd (2006). The standard account of spatial auto-correlation and its measurement is the book by Cliff and Ord (1973). More on local measures of spatial autocorrelation is provided by Anselin (1995) and Getis and Ord (1996). Fotheringham *et al.* (2002) provide a detailed account of GWR and the associated software.

⮕ The next chapter is concerned with methods for spatial interpolation—that is, methods for prediction of values at unsampled locations.

9

Spatial interpolation

9.1 Introduction

This chapter introduces a variety of approaches for the generation of surfaces, including topographic surfaces and other properties that can be treated as surfaces, such as precipitation amount or airborne pollutants. The term 'spatial interpolation' refers to the prediction of values at locations where no sample is available; a common objective in GIS contexts is to predict values on a regular grid using irregularly distributed point data, and this is the principal focus here.

There is a wide range of applications that depend on tools for predicting values on a regular grid from a set of samples irregularly distributed in space. An example is precipitation mapping. Precipitation amount may be measured using a set of rain gauges but if values are required elsewhere then it may be necessary to predict these values using the sample data. Another class of techniques that fall within the remit of this chapter are those that are used to transfer counts from one set of zones (e.g. census areas) to another set of zones or from zones to a grid. The term 'areal interpolation' describes such approaches.

9.2 Interpolation

Spatial interpolation approaches can be divided into those that are global and those that are local. Global approaches make simultaneous use of all sample data in the prediction process. Local approaches use only a subset of data (in a moving window) to make predictions. Another division into which spatial interpolation methods can be divided is exact and approximate methods. Exact methods 'honour' data locations—that is, observed values are not replaced and the predicted value at a location where there is a sample is the same as the sample value. With approximate methods, there is no guarantee that this is the case.

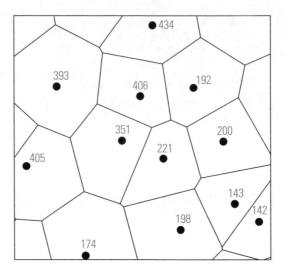

Figure 9.1 Thiessen polygons.

There are many approaches to spatial interpolation. A simple global approach is to fit a trend surface to all of the data, and values of the fitted surface can then be read off at any location, whether there is a sample at the location or not. Trend surface analysis entails fitting a plane or a curved surface through the data that represents the general trend of values (e.g. a 'gradual' increase in values from south to north). Trend surface analysis is simply multiple regression whereby the dependent variable is the variable of interest (e.g. precipitation) and the independent variables are the data coordinates or some function of them. Conceptually, the most simple interpolation method is Thiessen polygons. With this approach, the values assigned to unsampled locations are those of the nearest observation. Figure 9.1 gives an example of Thiessen polygons—the point value is assigned to the area which is closer to that point than any other.

The focus in this section is on methods that have a raster grid as their output. One interpolation framework that does not is triangulation. The output of triangulation is a triangulated irregular network (TIN), which is essentially a surface comprising triangular facets that connect the observations (see Heywood *et al.* (2006) for an example).

The following sections introduce five of the most widely used spatial interpolation methods: regression (introduced in Section 3.3, and expanded on in the previous chapter), TINs, inverse distance weighting (IDW), thin plate splines (TPS), and ordinary kriging (OK). There is a whole literature on TPS and geostatistical interpolation (kriging), and the accounts provided below are necessarily brief.

9.3 Triangulated irregular networks

The TIN is a vector-based representation of a surface. In essence it comprises a set of vertices joined by arcs, which together form triangles. TINs are more efficient than

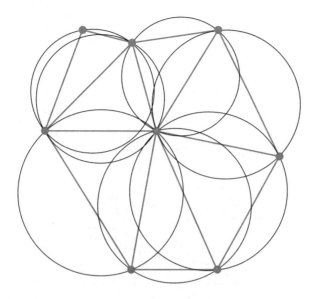

Figure 9.2 TIN subset with circumcircles superimposed.

raster-based digital elevation models (DEMs) as, in the latter case, elevations are stored at all locations on a regular grid while with TINs the sampling density can be varied as a function of the nature of the topography. Generally, marked breaks of slope are represented and there are usually more samples in areas with more variable elevations and fewer samples in relatively flat areas. Various approaches exist for the selection of observations with the objective of representing the surface as precisely as possible with the minimum of redundancy. One widely used approach to select points from a regular grid is the very important points (VIP) algorithm. With this approach, points are assigned a significance that is a function of the difference between each point and its neighbours. Following this procedure a specific number of the most significant points can be retained or points could be retained such that the loss of accuracy is minimized (Li *et al.*, 2004).

A TIN can be constructed from known point values using a process called Delaunay triangulation. Peuker *et al.* (1978) provide a detailed account of TINs. With Delaunay triangulation, the triangles are formed such that the circumcircle of each triangle contains no vertices other than those that make up the triangle. The circumcircles for a set of triangles are shown in Figure 9.2. As can be seen in the figure, the circumcircle runs through the three vertices that belong to a given triangle and in no case does a circumcircle for a given triangle contain any vertices other than those that belong to that triangle. With Delaunay triangulation, two vertices are connected if their Thiessen polygons share an edge, and this is illustrated in Figure 9.3. Delaunay triangulations can be derived from Thiessen polygons. Li *et al.* (2004) outline a variety of alternative approaches to selecting the starting point for, and conducting, Delaunay triangulation.

A TIN is illustrated using the Walker Lake sample of 470 point observations. These data come from the book by Isaaks and Srivastava (1989), and are provided through the AI-Geostats website (see http://www.ai-geostats.org/index.php?id=data).

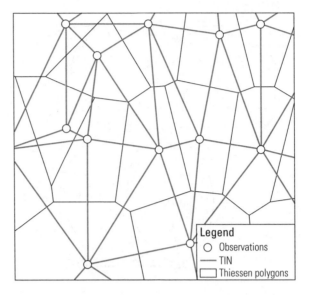

Figure 9.3 TIN subset with Thiessen polygons superimposed.

The illustrations here are based on the *V* variable derived from elevation data and described by Isaaks and Srivastava (1989). For the purposes of their example, Isaaks and Srivastava refer to the variable *V* as concentrations of some material in parts per million (ppm), here the variable is expressed in the same way, although the data are treated as elevation values for illustrative purposes. The data are used as they are preferentially sampled and allow the illustration of the TIN and some of its potential benefits in terms of smaller data storage requirements relative to an altitude matrix.

Figure 9.4 shows the point data, a raster grid generated with IDW (with an exponent of 2 and using eight nearest neighbours), and the edges of triangular facets derived using Delaunay triangulation. The TIN was generated using ArcGIS™ 3D Analyst.

Figure 9.5 shows a '2.5D' visualization of the TIN. In this case the 'elevations' are multiplied by 0.04 with respect to the *x* and *y* coordinate values, which has the effect of compressing the 'elevation' values. The ridge of large values running along the west of the region from north to south is apparent in Figure 9.5.

9.4 Regression for prediction

Regression (whether a global or local variant, like geographically weighted regression) can be used for spatial interpolation if values of the independent variable are available at all locations where predictions are required. In the previously outlined case of elevation and precipitation amount, if we have a DEM with elevation values at all locations in the study area then we can use the regression equation (either globally or locally) to predict precipitation amounts for all grid cell locations in the DEM (see, for example, Lloyd, 2005).

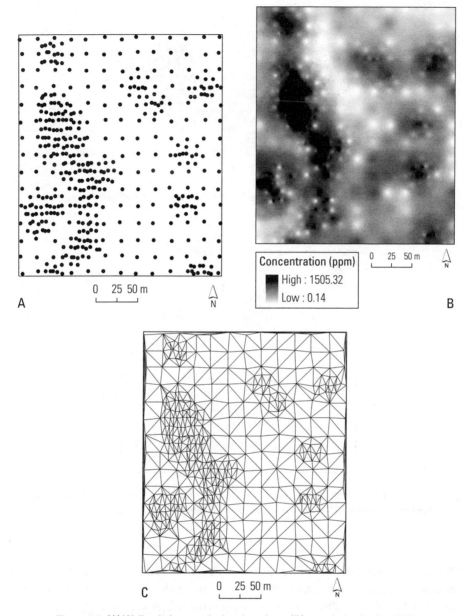

Figure 9.4 (A) Walker Lake sample data locations, (B) map derived using IDW with eight nearest neighbours, and (C) TIN.

9.4.1 Trend surface analysis

As summarized in Section 9.2, a trend surface is fitted using regression, but instead of regressing different variables, values are predicted using a regression of the dependent variable (e.g. elevation) against the coordinates or some function of them.

Figure 9.5 Shaded '2.5D' visualization of the TIN in Figure 9.4(C), viewed from the south east. Based on a *z* conversion factor of 0.04.

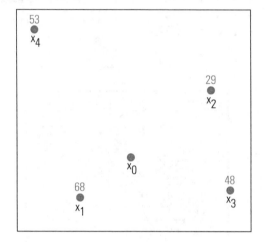

Figure 9.6 Location of prediction location and observations. Values and ID codes are given for the four samples and the value at location x_0 is treated as unknown.

For example, where the coordinates are given by x and y, the independent variables may be, for example, just x and y (this is called a first-order polynomial trend) or x, y, xy, x^2 and y^2 (this is called a second-order polynomial trend). Such approaches are useful for depicting general trends, but are unlikely to be of much practical use for direct interpolation.

9.5 Inverse distance weighting

Weighted moving averaging is a widely used approach to interpolation. A variety of different weighing functions have been used but IDW is the most common form in GIS.

IDW is an exact interpolator, so the predicted values at locations where there are observations are the same as at the observed values. The IDW predictor can be given as:

$$\hat{z}(\mathbf{x}_0) = \frac{\sum_{i=1}^{n} z(\mathbf{x}_i) \cdot d_{i0}^{-k}}{\sum_{i=1}^{n} d_{i0}^{-k}} \tag{9.1}$$

where the prediction is made at the location $\mathbf{x}_0$ as a function of the n neighbouring observations, $z(\mathbf{x}_i)$, $i = 1, \ldots, n$ (i.e. we feed only the n nearest neighbours into Equation 9.1), k is an exponent that determines the weight assigned to each of the observations, and d_{i0} is the distance by which the prediction location $\mathbf{x}_0$ and the observation location $\mathbf{x}_i$ are separated. As noted in Section 4.7, as the exponent becomes larger, the weight assigned to observations at large distances from the prediction location becomes smaller. Conversely, as was shown in Figure 4.6, for smaller values of the exponent, the weights are proportionally larger for more distant observations. The exponent is usually set to 2 (i.e. d_{i0}^{-2}) and the inverse squared distance (where $k = 2$) is obtained with $1/d^2$. The inverse square of a distance of 6481.996 m, is therefore $1/6481.996^2 = 0.00000002380$, as shown in Table 9.1.

In this section, a worked example of IDW is given using four observations, with the objective of predicting at another location (Figure 9.6 shows the data configuration). Since an observation is available at the prediction location $\mathbf{x}_0$, but it has been removed for the present purpose, it is possible to assess the accuracy of the predictions. The data are given in Table 9.1. The same data set is used to illustrate other interpolation methods in this chapter.

Following the IDW equation we first calculate the inverse square of the distances and then multiply these values by the value of the observations. Table 9.1 shows the inverse square distances and the observation values multiplied by the inverse square distances. The IDW prediction is given by (using the figures from Table 9.1) $0.000002526 / 0.00000004592 = 55.003$. The 'true' value at the prediction location is 61, so there is a prediction error of 5.997. In practice, assessment of prediction can be conducted using jackknifing or cross-validation. Jackknifing entails splitting the sample into two and using one set of data to make predictions at the locations represented by the second data set. The accuracy of these predictions can obviously be assessed directly. The basic idea of cross-validation was described in another context in Section 8.5.3. As detailed in that section, cross-validation entails removal of an observation, using the

Table 9.1 Precipitation (mm): IDW prediction using observations $\mathbf{x}_1$ to $\mathbf{x}_4$

i	x_i	y_i	$z(\mathbf{x}_i)$	d_{i0}	d_{i0}^{-2}	$z(\mathbf{x}_i) \cdot d_{i0}^{-2}$
1	292500	329100	68	6481.996	0.00000002380	0.000001618
2	305700	339700	29	10448.860	0.00000000916	0.000000266
3	307629	329826	48	10517.774	0.00000000904	0.000000434
4	287854	345702	53	15969.356	0.00000000392	0.000000208
				Sum	0.00000004592	0.000002526

remaining observations to predict the value of the removed value. The removed value is returned to the data set and the next observation in order is removed, after which the procedure is repeated for all observations. The prediction errors can then be assessed. The accuracy of prediction is commonly assessed using summaries such as the mean error, the mean absolute error, and the root mean square error (RMSE).

In terms of selecting a data subset for interpolation, several common strategies exist. The n (four in the example) nearest neighbours to a prediction location could be selected. Alternatively, all observations within a specified distance of the prediction location could be used. Another strategy is to divide the search neighbourhood into quadrants, for example north-east, south-east, north-west, and south-west of the pre-diction location. The weights could then be scaled according to the number of obser-vations in each quadrant and this would help to overcome the effect of clustering of observations in particular areas.

IDW was used to generate a map of precipitation amount in July 2006 in Northern Ireland using the 16 nearest neighbours to each cell of the prediction grid. The data locations are shown in Figure 8.7 and the IDW-derived map is shown in Figure 9.7. The map in Figure 9.7 is very smooth in appearance and there are clear clusters of values around the sample locations—this is a common feature of maps generated using IDW. With IDW, there tend to be clusters of similar values around data points (see Lloyd (2005) for another example).

IDW is rapid and easy to implement, although it often performs less well than more sophisticated approaches (e.g. see Lloyd, 2005).

9.6 Thin plate splines

TPS are, like IDW, a very widely used approach to spatial interpolation. TPS functions are available in ArcGIS™, the GRASS GIS (Neteler and Mitásová, 2007), and in other software packages. TPS can be viewed as surfaces that are fitted to some local subset of the data. The spline can be fitted exactly to the data points or it can be smoothed—that is, if the spline is not forced to fit to the data points the resulting surface can be made smoother than if the surface runs through every point. In effect, the thin plate smooth-ing spline generated map is a map of local weighted averages. With TPS, the aim is to obtain a prediction of the unknown value $\hat{z}(\mathbf{x}_0)$ with a smooth function g by minimizing (as defined below):

$$\sum_{i=1}^{n}(z(\mathbf{x}_i) - g(\mathbf{x}_i))^2 + \rho J_m(g) \tag{9.2}$$

where $J_m(g)$ is a measure of the roughness of the spline function, m is the degree of the polynomial used in the model, and ρ is the smoothing parameter. We seek to find the function g so that it is as close as possible to the observations (indicated by $z(\mathbf{x}_i)$), with the smoothing function determining if the fit of the function to the observations is

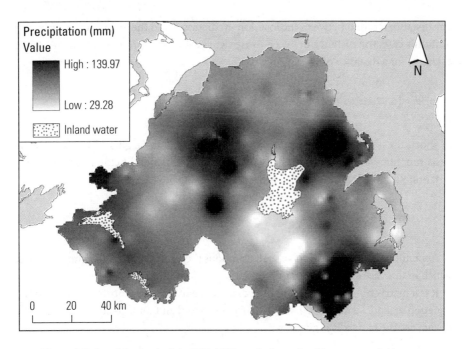

Precipitation (mm)
Value

High : 139.97

Low : 29.28

Inland water

0   20   40 km

N

Figure 9.7 Precipitation in July 2006: IDW prediction using 16 nearest neighbours.

exact or approximate. If the smoothing parameter is zero then the spline passes through the data (it is an exact interpolator); if the value is greater than zero then the spline is not forced to fit to the data (it is an approximate interpolator). If the data are 'noisy' (e.g. there is notable measurement error), use of a smoothing parameter may be desirable. Large values of the smoothing parameter result in smoother maps. The TPS function g is made up of two parts (as defined after the equation):

$$g(\mathbf{x}) = a_0 + a_1 x + a_2 y + \sum_{i=1}^{n} \lambda_i R(\mathbf{x} - \mathbf{x}_i) \qquad (9.3)$$

The aim is to find values of a_0, a_1, a_2, and λ_i to make a prediction, as we know the other terms (which will be detailed below) in advance. The left-hand side of the equation (i.e. $a_0 + a_1 x + a_2 y$) indicates the local trend in the data. The introduction to this chapter, as well as Section 9.4, mentioned trend surface analysis, in which a surface (either a flat plane or a more complex surface) is fitted to the data in the same way that a line is fitted to a scatter plot using regression. In the case of splines, a surface of this kind is fitted to the data, but only to some local subset (e.g. the 16 nearest neighbours to the prediction location). In the same way as a slope value is found in (bivariate) linear regression and multiplied by the independent variable, we must find values for a_1 and a_2, and these will be multiplied by x and y. The term $R(\mathbf{x} - \mathbf{x}_i)$ is called a basis function and for TPS it is given by:

$$d_i^2 \log d_i \qquad (9.4)$$

The distance d is that between the location prediction $\mathbf{x}$ and the location $\mathbf{x}_i$, so $R(\mathbf{x}-\mathbf{x}_i)$ is, in this case, the distance between those two locations fed into Equation 9.4. As an example, for a distance of 16.9292646 units:

$$d_i^2 \log d_i = 16.9292646^2 \log 16.9292646 = 352.128$$

In short, the TPS function comprises the local trend and weights (λ_i) by which the basis function values are multiplied (the process is illustrated below). Using matrix notation, the coefficients a_k and λ_i are the solution of:

$$\mathbf{R}\lambda = \mathbf{z} \tag{9.5}$$

Appendix F shows how to solve such equations (i.e. how to find the unknown values, which in this case are the coefficients a_k and λ_i).

$\mathbf{R}$ is a matrix obtained by feeding the distances between local observations into the equation $d^2 \log d$. As above, a given distance is squared and multiplied by the log of the distances. The matrix $\mathbf{R}$ is given by:

$$\mathbf{R} = \begin{bmatrix}
R(\mathbf{x}_1 - \mathbf{x}_1) & \cdots & R(\mathbf{x}_1 - \mathbf{x}_n) & 1 & x_1 & y_1 \\
\vdots & \vdots & \vdots & \vdots & \vdots & \vdots \\
R(\mathbf{x}_n - \mathbf{x}_1) & \cdots & R(\mathbf{x}_n - \mathbf{x}_n) & 1 & x_n & y_n \\
1 & \cdots & 1 & 0 & 0 & 0 \\
x_1 & \cdots & x_n & 0 & 0 & 0 \\
y_1 & \cdots & y_n & 0 & 0 & 0
\end{bmatrix}$$

λ are the TPS weights and $\mathbf{z}$ are the observations:

$$\lambda = \begin{bmatrix} \lambda_1 \\ \vdots \\ \lambda_n \\ a_0 \\ a_1 \\ a_2 \end{bmatrix} \qquad \mathbf{z} = \begin{bmatrix} z(\mathbf{x}_1) \\ \vdots \\ z(\mathbf{x}_n) \\ 0 \\ 0 \\ 0 \end{bmatrix}$$

The matrix $\mathbf{R}$ has the weights and three rows and columns corresponding to a constant trend component (the 1s) and the x and y coordinates of each location, the vector (a matrix with only one row or column) λ has three extra rows, which are values of a_k for the constant (a_0) and for the x and y coordinates of each location (i.e. a_1 and a_2), and the vector $\mathbf{z}$ includes three zeros, corresponding to the three trend components.

To obtain the TPS weights (λ_i) and the values of a_0, a_1, and a_2, the inverse (see Section 3.3 and Appendices E and F for a discussion about matrix inversion) of the matrix $\mathbf{R}$ is multiplied by the vector of data values, $\mathbf{z}$:

$$\lambda = \mathbf{R}^{-1}\mathbf{z}$$

Using the same data as for the IDW example in Section 9.5, following Equation 9.5 the TPS system is given as:

$$
\begin{bmatrix}
0 & 352.128 & 270.779 & 367.512 & 1 & 292.500 & 329.100 \\
352.128 & 0 & 101.483 & 451.925 & 1 & 305.700 & 339.700 \\
270.779 & 101.483 & 0 & 902.999 & 1 & 307.629 & 329.826 \\
367.512 & 451.925 & 902.999 & 0 & 1 & 287.854 & 345.702 \\
1 & 1 & 1 & 1 & 0 & 0 & 0 \\
292.500 & 305.700 & 307.629 & 287.854 & 0 & 0 & 0 \\
329.100 & 339.700 & 329.826 & 345.702 & 0 & 0 & 0
\end{bmatrix}
\times
\begin{bmatrix}
\lambda_1 \\ \lambda_2 \\ \lambda_3 \\ \lambda_4 \\ a_0 \\ a_1 \\ a_2
\end{bmatrix}
=
\begin{bmatrix}
68 \\ 29 \\ 48 \\ 53 \\ 0 \\ 0 \\ 0
\end{bmatrix}
$$

Note that the distances were divided by 1000 prior to calculating the values for $\mathbf{R}$. This gives the same results but reduces the size of the values obtained for $R(\mathbf{x}-\mathbf{x}_i)$ and makes the process more manageable. The diagonals in the matrix $\mathbf{R}$ are all 0 and they indicate the distance between an observation and itself (obviously 0); where a smoothing parameter is used (and the spline function is not forced to fit to the data), the smoothing parameter value is added to the diagonals (for this example, the first four 0 components, reading from the left), as described by Lloyd (2006).

Solving the TPS system, the weights are as follows: $\lambda_1 = -0.0319$, $\lambda_2 = -0.0493$, $\lambda_3 = 0.0520$, $\lambda_4 = 0.0292$, $a_0 = 1078.474$, $a_1 = -1.5906$, and $a_2 = -1.6794$.

We then put the distances between each observation and the prediction location into the equation $d^2 \log d$. For each observation this gives the following values: $\mathbf{x}_1 = 34.105$, $\mathbf{x}_2 = 111.261$, $\mathbf{x}_3 = 113.049$, and $\mathbf{x}_4 = 306.863$.

The predicted value is then given by multiplying the weights (λ_i) by the basis function values ($R(\mathbf{x}-\mathbf{x}_i)$), adding a_0 (the constant, note that the intercept is also called the constant), and multiplying the trend coefficients (a_1 and a_2) by the coordinates (x and y). In this case this leads to: $(34.105 \times -0.0319) + (111.261 \times -0.0493) + (113.049 \times 0.0520) + (306.863 \times 0.0292) + 1078.474 + (297.624 \times -1.5906) + (333.070 \times -1.6794) = 60.569$.

The 'true' value is 61 and the TPS prediction error is smaller than the IDW prediction error (with an IDW prediction of 55.003; see Section 9.5).

Figure 9.8 shows a map of precipitation amount generated using TPS with the same data used to illustrate the application of IDW (see Figure 9.7). Lloyd (2006) gives a summary account of variants of the TPS approach, which may provide more robust and more accurate predictions than standard TPS in some circumstances.

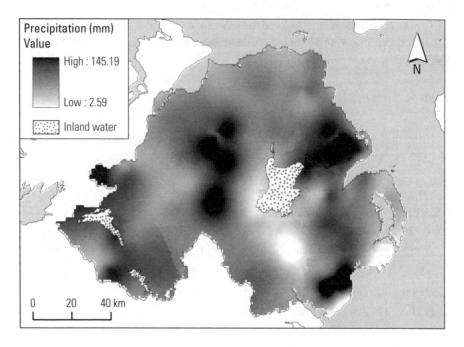

Figure 9.8 Precipitation in July 2006: TPS prediction using 16 nearest neighbours.

9.7 Ordinary kriging

Ordinary kriging is one of a family of kriging methods. Kriging falls within the remit of a field known as geostatistics. Burrough and McDonnell (1998) provide a short introduction to geostatistics in the context of GIS. The basis of geostatistics is the theory of regionalized variables. Geostatistics entails a conceptual division of spatial variation (at a location $\mathbf{x}$) into two distinct parts: a deterministic component ($\mu(\mathbf{x})$) (representing 'gradual' change over the study area) and a stochastic (or 'random') component ($R(\mathbf{x})$):

$$Z(\mathbf{x}) = \mu(\mathbf{x}) + R(\mathbf{x}) \tag{9.6}$$

This is termed a 'random function' (RF) model. The random part reflects our uncertainty about spatial variables—what seems random to us is a function of a multiplicity of factors that may be impossible to model directly (and this does not mean that we really think variation is random; Isaaks and Srivastava, 1989). In geostatistics, a spatially referenced variable, $z(\mathbf{x})$, is treated as an outcome of an RF, $Z(\mathbf{x})$. In other words, we effectively consider an observation to have been generated by the RF model and this gives us a framework to work with these data. A realization of an RF is called a regionalized variable (ReV; i.e. an observation). The theory of regionalized variables (Matheron, 1971) is the fundamental framework on which geostatistics is based.

The discussion on first- and second-order effects in Section 7.1 is relevant in this context as it deals with the distinction between variation in the mean and spatial dependence. This section also has links with the introduction to TPS, whereby the TPS function (Equation 9.3) is shown to comprise a trend component and the component (here termed the 'random' part) modelled by the basis function $R(\mathbf{x} - \mathbf{x}_i)$ (in the present case, the variogram is used instead to model this component, as described below).

In practical terms, as we estimate parameters, namely the mean and variance, of a distribution, we estimate parameters of the RF model using the data. These parameters, like the mean and variance, summarize the variable. The mean and variance of a distribution are useful only if the distribution is approximately normal and, similarly, the parameters of the RF model are only meaningful in certain conditions. Where the properties of the variable of interest are the same, or at least similar in some sense, across the region of interest we can employ a stationary model. In other words, we can use the same model parameters at all locations. If the properties of the variable are clearly spatially variable then a standard RF model may not be appropriate. There are different degrees of stationarity, but for present purposes we will only consider one, intrinsic stationarity. There are two requirements of intrinsic stationarity. Firstly, the mean is constant across the region of interest. In other words, the expected value of the variable does not depend on the location, $\mathbf{x}$:

$$E\{Z(\mathbf{x})\} = \mu(\mathbf{x}) \text{ for all } \mathbf{x} \tag{9.7}$$

The mean is therefore assumed to be the same for all locations. Secondly, the expected squared difference between paired RFs (i.e. the observations) (summarized by the variogram, $\gamma(\mathbf{h})$) should depend only on the separation distance and direction (the lag $\mathbf{h}$) between the RFs and not on the location of the RFs:

$$\gamma(\mathbf{h}) = \frac{1}{2} E[\{Z(\mathbf{x}) - Z(\mathbf{x} + \mathbf{h})\}^2] \text{ for all } \mathbf{h} \tag{9.8}$$

where $\mathbf{x} + \mathbf{h}$ indicates a distance (and direction) $\mathbf{h}$ from location $\mathbf{x}$.

In terms of the data, the expected semivariance should be the same for all observations separated by a particular lag, irrespective of where the paired observations are located. In practical terms, the geostatistical approach can be applied irrespective of these conditions, but the results will clearly be suboptimal if the data depart markedly from the conditions. In some cases the mean is allowed to vary from place to place, but stay constant within a moving window. This is known as quasi stationarity (Webster and Oliver, 2007).

9.7.1 Variogram

Analysis of the degree to which values differ according to how far apart they are can be conducted by computing the variogram (or semivariogram). With reference to the variogram, the term 'lag' is used to describe the distance and direction by which observations are separated. For example, two observations may be 5 km apart and one may be directly north of the other. In simple terms, the variogram is estimated by calculating the squared differences between all the available paired observations and obtaining half the average for all observations separated by that lag (or within a lag tolerance (e.g. 5±2.5 km) where the observations are not on a regular grid). Semi-variance refers to half the squared difference between data values. An example of variogram estimation is given below. Figure 9.9 gives a simple example of a transect along which observations have been made at regular intervals. Lags (**h**) of 1 and 2 are indicated. In this case, therefore, half the average squared difference between observations separated by a lag of 1 is calculated and the process is repeated for a lag of 2 and so on. In many cases the distance between observations will not be regular, so ranges of distances are grouped. The selection of the bin size (e.g. 0–5 km, >5–10 km, >10–15 km, … or 0–10 km, >10–20 km, …) is important. Smaller bin sizes will result in more noisy variograms, while a bin size that is too large will smooth out too much spatial structure and it will not be possible to capture spatial variation of interest. In other words, the plotted values in a variogram with too small a bin size will appear to be widely scattered, while the values in a variogram with a larger bin size will tend to be more similar to neighbouring values on the plot. Finding an appropriate bin size is important in characterizing spatial structure and in guiding the selection and fitting of a model, as detailed below.

The variogram can be estimated for different directions to enable the identification of directional variation (anisotropy). In other words, rather than consider all observations 5 km from a given observation, we may consider only observations that are directly north or south (for example) of the observation of interest within a particular angular tolerance (e.g. north or south ±45 degrees). An example of an anisotropic phenomenon is temperature—Hudson and Wackernagel (1994) showed that average January temperature in Scotland decreased systematically from west to east (a function of the warming effect of the Gulf Stream), but there was no systematic trend in a north to south direction.

In summary, the variogram characterizes the degree of difference in values as a function of the distance by which they are separated. The experimental variogram, $\hat{\gamma}(\mathbf{h})$, relates semivariances to distances (and directions)—it has distance and direction (the lag) on the x axis and semivariance on the y axis. If a property is spatially auto-correlated, we would expect the semivariance to increase as the distance between observations increases.

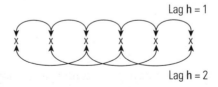

Figure 9.9 Transect with paired points selected for lags of 1 and 2 units.

As an example, if we take a distance range of 1000 to 2000 m and there are 346 pairs of observations separated by a distance within that band then $p(\mathbf{h})$, the number of paired observations, is 346. Note that for each pair (e.g. observations 23 and 37), the semivariance is calculated twice: once with respect to the first location and once with respect to the second. We then calculate the squared difference between each of these paired values. The first value in each pair is given by $z(\mathbf{x}_i)$ and the value separated from it by the specified lag $\mathbf{h}$ (in this example the distance is 1000 to 2000 m and we are concerned with all directions) is given by $z(\mathbf{x}_i + \mathbf{h})$. Their squared difference is therefore given by $\{z(\mathbf{x}_i) - z(\mathbf{x}_i + \mathbf{h})\}^2$. The summed values are then divided by two, hence the term 'semivariance'. Putting this together, the experimental variogram for lag $\mathbf{h}$ is computed from:

$$\hat{\gamma}(\mathbf{h}) = \frac{1}{2p(\mathbf{h})} \sum_{i=1}^{p(\mathbf{h})} \{z(\mathbf{x}_i) - z(\mathbf{x}_i + \mathbf{h})\}^2 \tag{9.9}$$

As an example, if our lag is 5 ± 2.5 km (i.e. 2.5 to 7.5 km) and we have two values separated by 6.2 km, then these paired observations qualify and we compute the squared difference. If the two values are 26.2 and 43.3 then their squared difference is:

$$\{z(\mathbf{x}_i) - z(\mathbf{x}_i + \mathbf{h})\}^2 = \{26.3 - 43.4\}^2 = \{-17.1\}^2 = 292.41$$

In the same way we compute the squared difference for all other pairs separated by 2.5 to 7.5 km and at each stage add the computed value to the previous values computed for that lag. Once this is done, we multiply the summed values by $^1/(2p(\mathbf{h}))$.

Figure 9.10 gives an example of an experimental variogram estimated from the precipitation data introduced in Section 9.5. The Gstat software (Pebesma and Wesseling, 1998; Pebesma, 2004) was used to estimate the variogram. The lags are 0–5000 m, 5000–10,000 m and so on in groups of 5000 up to 60,000 m. In this case, data values are

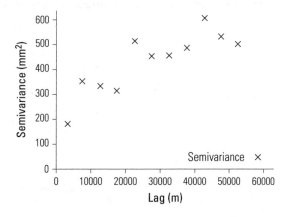

Figure 9.10 Omnidirectional variogram of July 2006 precipitation amount in Northern Ireland.

compared irrespective of the direction in which they are aligned—that is, whether they are aligned (approximately or absolutely) on a line north–south or east–west, etc. of one another is irrelevant. A variogram computed from data in all directions is termed 'omnidirectional'.

In Figure 9.10, the semivariance values tend to be smaller for small lags and they generally increase with an increase in lag size until perhaps 25,000 m, where the values tend to level out (this is demonstrated below). This indicates that values are positively spatially autocorrelated up to approximately this distance. At distances larger than this, there is no spatial structure. The variogram provides a useful means of summarizing how values change with separation distance. Using topography as an example, data representing a 'smooth' surface like a flood plain will have a very different variogram to data representing a 'rough' surface like a mountain range.

A mathematical model may be fitted to the experimental variogram and the coefficients of this model can be used for spatial prediction using kriging or for conditional simulation (defined below). A model can be fitted 'by eye' or by using some fitting procedure such as ordinary least squares (see Sections 3.3 and 8.5, and Appendix E) or weighted least squares. A model is usually selected from one of a set of 'authorized' models. Webster and Oliver (2007) provide a review of some of the most widely used authorized models.

There are two principal classes of variogram model. Transitive (bounded) models have a sill (finite variance)—that is, the variogram levels out as it reaches a particular lag. Unbounded models do not reach an upper bound. Figure 9.11 shows the components of a bounded variogram model. These will be defined and then practical examples given. The nugget effect, c_0, represents unresolved variation (a mixture of spatial variation at a finer scale than the sample spacing and measurement error). The structured component, c, represents the spatially correlated variation. The sill (or sill variance), $c_0 + c$, is the *a priori* variance. The range, a, represents the scale (or frequency) of spatial variation. For example, if a region is mountainous and elevation varies markedly over quite small distances, then the elevation can be said to have a high frequency of spatial variation (a short range), while if the elevation is quite similar over much of the area (e.g. it is a river flood plain) and varies markedly only at the extremes of the

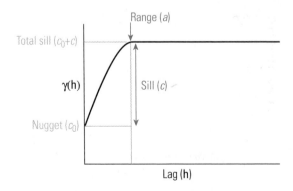

Figure 9.11 Bounded variogram model.

site (i.e. at large separation distances), then the elevation can be said to have a low frequency of spatial variation (a long range).

As noted above, there are many different models that can be fitted to variograms. The variogram illustrated above was fitted with a nugget effect and a spherical component. The nugget effect (nugget variance) is given as:

$$\gamma(h) = \begin{cases} 0 & \text{if } h = 0 \\ c_0 & \text{if } h > 0 \end{cases} \qquad (9.10)$$

In other words, the modelled semivariance has a value of 0 for a lag of 0, but is equal to c_0 for all positive values of the lag. In Figure 9.11, the nugget effect is indicated on the y axis of the graph.

The spherical model, a bounded model (i.e. it reaches a sill) is defined as:

$$\gamma(h) = \begin{cases} c \cdot [1.5 \frac{h}{a} - 0.5 \left(\frac{h}{a}\right)^3] & \text{if } h \leq a \\ c & \text{if } h > a \end{cases} \qquad (9.11)$$

where c is called, as noted above, the structured component. In other words, the modelled semivariance is computed using the top line for all lag values up to and including the range. For lag values larger than the range the modelled semivariance is equal to c. Authorized models may be used in combination where a single model is insufficient to properly represent the form of the variogram. For example, if the spatial structure is complex and does not simply increase and level out (as in the example in Figure 9.11) then models may be combined to take this complexity into account (e.g. a model could comprise a nugget effect and two spherical components, thus there would be two breaks of slope, rather than just one). Figure 9.12 shows an omnidirectional variogram of July 2006 precipitation amount in Northern Ireland with a fitted model comprising

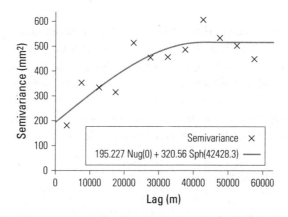

Figure 9.12 Omnidirectional variogram of July 2006 precipitation amount in Northern Ireland, with fitted model.

a nugget effect (with a value of 195.227) and a spherical component (with a structured component of 320.56 and a range of 42428.3 m).

The model fitted to the variogram can be used to determine the weights assigned to observations using a geostatistical prediction procedure (or family of procedures) called kriging. In kriging, the variogram model is used to obtain values of the semi-variance for the lags by which observations are separated and for the lags that separate the prediction location from the observation.

For the variogram model in Figure 9.12 there is a nugget effect and a spherical component. Combining Equations 9.10 and 9.11 this gives:

$$\gamma(h) = \begin{cases} c_0 + c \cdot [1.5\frac{h}{a} - 0.5(\frac{h}{a})^3] & \text{if } h \le a \\ c_0 + c & \text{if } h > a \end{cases}$$

For a lag of 6481.996 m (which is less than the range value of 42428.3 m) the modelled semivariance is obtained from:

$$\gamma(6481.996) = 195.227 + 320.560 \cdot \left[1.5\frac{6481.996}{42428.3} - 0.5\left(\frac{6481.996}{42428.3}\right)^3\right] = 268.116 \text{ m}$$

This will be confirmed by examining Figure 9.12 and reading upwards from a lag of 6482 m to the variogram model and then left to read off the semivariance value. Mulla (1988) used variograms, along with other measures, to characterize landforms. As noted previously, in terms of landforms, a mountainous area would have a short range, since there are large changes in elevation over small distances. In contrast, a river flood plain would have a long range as elevations tend to be similar over quite large distances. The variogram is, therefore, a useful tool for measuring the scale of spatial variation in a property. Prediction using kriging, which makes use of the variogram model, is the subject of the following section.

9.7.2 Kriging

There are many varieties of kriging. Its simplest form is called simple kriging (SK). To use SK it is necessary to know the mean of the property of interest and this must be constant across the region of interest. In practice this is rarely the case. The most widely used variant of kriging, ordinary kriging (OK), allows the mean to vary and the mean is estimated for each prediction neighbourhood. OK predictions are weighted averages of the n available data (i.e. the predictions are based on the n nearest neighbours of the prediction location). The OK prediction, $\hat{z}(\mathbf{x}_0)$, is defined as:

$$\hat{z}(\mathbf{x}_0) = \sum_{i=1}^{n} \lambda_i z(\mathbf{x}_i) \tag{9.12}$$

with the constraint that the weights, λ_i, sum to 1 (this is to ensure an unbiased prediction):

$$\sum_{i=1}^{n} \lambda_i = 1 \tag{9.13}$$

The objective of the kriging system is to find appropriate weights by which the available observations will be multiplied before summing them to obtain the predicted value. These weights are determined using the coefficients of a model fitted to the variogram (or another function such as the covariance function). This is in contrast to IDW (Section 9.5), where the weights are selected arbitrarily (i.e. not using information about the spatial variation in the data).

The weights are obtained by solving (i.e. finding the values of unknown coefficients in) the OK system:

$$\begin{cases} \sum_{j=1}^{n} \lambda_j \gamma(\mathbf{x}_i - \mathbf{x}_j) + \psi = \gamma(\mathbf{x}_i - \mathbf{x}_0) & i = 1,...,n \\ \sum_{j=1}^{n} \lambda_j = 1 \end{cases} \tag{9.14}$$

where ψ is the Lagrange multiplier. This equation may seem at first sight complicated. In words, it says that the sum of the weights multiplied by the modelled semivariance for the lag separating locations $\mathbf{x}_i$ and $\mathbf{x}_j$ plus the Lagrange multiplier equals the semivariance between locations $\mathbf{x}_i$ and the prediction location $\mathbf{x}_0$ with the constraint that the weights must sum to 1. The way we find the weights and the Lagrange multiplier is outlined below.

Computing the weights and a value of the Lagrange multiplier, ψ, allows us to obtain the prediction variance of OK, a by-product of OK, which can be given as:

$$\hat{\sigma}_{OK}^2 = \sum_{i=1}^{n} \lambda_i \gamma(\mathbf{x}_i - \mathbf{x}_0) + \psi \tag{9.15}$$

The kriging variance is a measure of confidence in predictions and is a function of the form of the variogram, the sample configuration, and the sample support (the area over which an observation is made, which may be approximated as a point or may be an area) (Journel and Huijbregts, 1978). If the variogram model range is short then the kriging variance will increase markedly with distance from the nearest samples. There are two varieties of OK: punctual OK and block OK. With punctual OK the predictions cover the same area (the support, V) as the observations. In block OK, the predictions are made to a larger support than the observations (e.g. prediction from points to areas of 2 m by 2 m). The system presented here is for the more commonly used form, punctual OK.

Returning to Equation 9.14, using matrix notation, the OK system can be written as:

$$\mathbf{K}\lambda = \mathbf{k} \tag{9.16}$$

where $\mathbf{K}$ is the $n+1 \times n+1$ (with n nearest neighbours used for prediction) matrix of semivariances between each of the observations:

$$\mathbf{K} = \begin{bmatrix} \gamma(\mathbf{x}_1 - \mathbf{x}_1) & \cdots & \gamma(\mathbf{x}_1 - \mathbf{x}_n) & 1 \\ \vdots & \vdots & \vdots & \vdots \\ \gamma(\mathbf{x}_n - \mathbf{x}_1) & \cdots & \gamma(\mathbf{x}_n - \mathbf{x}_n) & 1 \\ 1 & \cdots & 1 & 0 \end{bmatrix}$$

λ are the OK weights and $\mathbf{k}$ are semivariances for the observations to the prediction location (with one placed in the bottom position):

$$\lambda = \begin{bmatrix} \lambda_1 \\ \vdots \\ \lambda_n \\ \psi \end{bmatrix} \qquad \mathbf{k} = \begin{bmatrix} \gamma(\mathbf{x}_1 - \mathbf{x}_0) \\ \vdots \\ \gamma(\mathbf{x}_n - \mathbf{x}_0) \\ 1 \end{bmatrix}$$

To obtain the OK weights, the inverse of the data semivariance matrix is multiplied by the vector of data to prediction semivariances:

$$\lambda = \mathbf{K}^{-1}\mathbf{k} \tag{9.17}$$

The OK variance is then obtained from:

$$\sigma^2_{OK} = \mathbf{k}^T \lambda \tag{9.18}$$

Using the same data as for the example in Sections 9.5 (IDW) and 9.6 (TPS), the OK system is given as:

$$\begin{bmatrix} 0 & 376.905 & 359.589 & 379.853 & 1 \\ 376.905 & 0 & 307.108 & 394.601 & 1 \\ 359.589 & 307.108 & 0 & 448.401 & 1 \\ 379.853 & 394.601 & 448.401 & 0 & 1 \\ 1 & 1 & 1 & 1 & 0 \end{bmatrix} \times \begin{bmatrix} \lambda_1 \\ \lambda_2 \\ \lambda_3 \\ \lambda_4 \\ \psi \end{bmatrix} = \begin{bmatrix} 268.116 \\ 311.250 \\ 311.983 \\ 367.662 \\ 1 \end{bmatrix}$$

Note that the semivariance between a given location and itself is set to 0.

This account does not show how the weights are obtained. To see how this is done (i.e. to see how the OK system is solved) go to Appendix F, where the same example is given (but exactly how the system is solved is shown).

Solving the OK system, the weights are as follows: $\lambda_1 = 0.368$, $\lambda_2 = 0.227$, $\lambda_3 = 0.234$, $\lambda_4 = 0.171$, and $\psi = 33.332$.

The predicted value is then given by: $(0.368 \times 68) + (0.227 \times 29) + (0.234 \times 48) + (0.171 \times 53) = 51.889$.

The kriging variance is given by: $(0.368 \times 268.116) + (0.227 \times 311.250) + (0.234 \times 311.983) + (0.171 \times 367.662) + (33.332 \times 1) = 338.537$.

The kriging variance is a useful by-product which, as detailed above, provides a guide to uncertainty in predicted values.

The 'true' precipitation amount value is 61 mm, so there is a prediction error of 9.111. In this case, the IDW prediction of 55.003 (see Section 9.5) is closer to the true value, as is the TPS prediction of 60.569 (see Section 9.6). There is, of course, no guarantee that OK will provide more accurate predictions than IDW, despite the use of arbitrary weights in the latter case, but many real-world case studies have shown how techniques like OK often provide an increase in prediction accuracy over simpler methods like IDW (see Lloyd (2005) for an example). Lloyd (2006) and Chang (2008) provide worked examples of IDW, TPS, OK, and other approaches.

Figure 9.13 shows a map of precipitation in July 2006 generated using OK with 16 nearest neighbours.

Comparison of Figure 9.13 with Figures 9.7 (IDW derived map) and 9.8 (TPS derived map) shows quite large differences in the range of values. This demonstrates the large variations that can result from the application of different interpolation procedures. This issue is discussed further below.

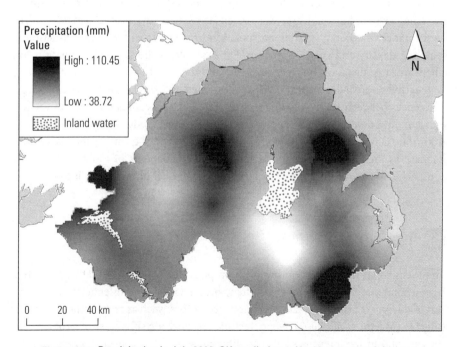

Figure 9.13 Precipitation in July 2006: OK prediction using 16 nearest neighbours.

9.7.3 Cokriging

There are several other forms of kriging; cokriging, for example, allows the integration of information about secondary variables. In cases where we have a secondary variable (or variables) that is cross-correlated with the primary variable, both (or all) variables may be used simultaneously to make predictions using cokriging. With cokriging, the variograms (which can be termed 'autovariograms') of both (or all) variables and the cross-variogram (describing the spatial dependence between the two variables) must be estimated and models fitted to all of these. Cokriging is based on the linear model of coregionalization (see Webster and Oliver, 2007). For cokriging to be beneficial, the secondary variable should be cheaper to obtain or more readily available than the primary variable (i.e. the variable that will be mapped). If the variables are strongly related linearly then cokriging may provide more accurate predictions than OK.

9.8 Other approaches and issues

There are many other widely used spatial interpolation approaches in addition to variants of TPS and kriging. Several approaches are summarized by Mitás and Mitásová (1999). Specialist routines have been written for some applications. For example, the routine of Hutchinson (1989) is used specifically for generating DEMs. Clearly, the selection of an interpolation method impacts on the final results, and researchers have assessed variations in results following application of different interpolation methods (see Chapter 10 of Burrough and McDonnell (1998) for a review of related topics). The performance of different spatial interpolation procedures will vary as a function of sampling density and spatial variation. For example, if the sampling density is low (there are large distances between samples) and there is short-range spatial variation (values differ a great deal over short distances), then we would expect there to be larger differences between results obtained using different procedures than in cases where the sampling density is high and there is long-range spatial variation. Lloyd and Atkinson (2002) show how differences in predictions (in terms of their accuracy) increase as the sampling density decreases, and the benefits of more sophisticated approaches are shown to be more apparent where the sampling density is low.

9.9 Areal interpolation

The focus so far in this section has been on point interpolation—that is, prediction from a point sample to a regular grid. Often, there is a need to transfer between different sets of zones or transfer, for example, counts from zones (such as census reporting areas) to grids (Martin *et al.*, 2002). Many techniques exist for solving such problems. In the case of transferring values between different sets of zones, overlay procedures (as detailed in Chapter 5) provide a partial solution. Counts could be reassigned to new zones

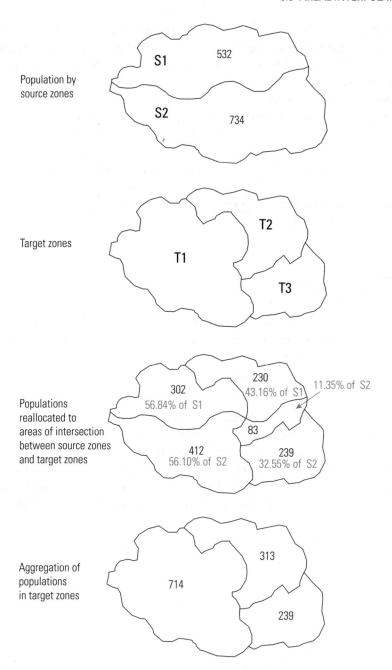

Figure 9.14 Areal weighting example.

according to the size of the overlapping areas between the old zone set and the new zone set. Such an approach is called areal weighting and an example is given in Figure 9.14.

As a particular example, zone T1 covers 56.84% of the area of zone S1. Given this information, the expected population of the area of overlap (i.e. intersection) between zones T1 and S1 is 56.84% of the zone S1 population of 532, thus $(^{532}/100) \times 56.84 = 302$ (when rounded to a whole number). Once the populations of each of the areas of intersection have been obtained they can be summed within the target zones, as shown in the bottom part of Figure 9.14.

More sophisticated approaches to areal interpolation exist (see Lloyd (2006) for a summary). The main focus in this chapter is on the generation of surfaces rather than zones, and so of more immediate relevance here are approaches such as the pycnophylactic reallocation method of Tobler (1979) or the population surface modelling procedure of Martin (1989). Both of these approaches allow the transfer of zonal counts to regular grids. One benefit of such approaches is to enable direct comparison of values for different time periods even when the original zonal systems used at different periods are quite different.

9.10 Case studies

The following two case studies are based on the data introduced in Section 8.7. These case studies demonstrate (1) estimation of the variogram and (2) spatial interpolation using IDW, TPS, and OK.

9.10.1 Variogram estimation

Figure 9.15 shows an experimental variogram computed from precipitation data. Recall that the variogram tells us how different observations tend to be a function of how far apart the observations are. In this case, the semivariance values increase with increased distance and they level out at a distance of perhaps 75 km. A model can be fitted to the variogram, as detailed in Section 9.7.1, and the model used to inform prediction of precipitation amount at locations where there are no measurements

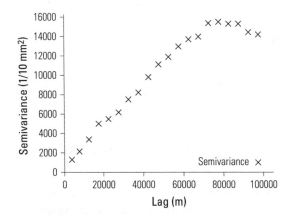

Figure 9.15 Experimental variogram computed from precipitation data for 8 May 1986 in Switzerland.

(see Section 9.7.2 for a discussion about this). The use of a fitted model for interpolation is detailed in the next subsection. In this case, the geostatistical software Gstat (Pebesma and Wesseling, 1998; Pebesma, 2004) was used for the analysis; ArcGIS™ Geostatistical Analyst could also be used (although the variogram will appear slightly different given the way the semivariances are computed in that package).

9.10.2 Interpolation

Precipitation amounts were predicted on a regular grid using the IDW, TPS, and OK functions of ArcGIS™ Geostatistical Analyst. For OK, the variogram was estimated using a lag size of 5000 and 20 lags, with the model (nugget effect=403.9, spherical component (partial sill)=14689, and range=90653.3 m) fitted using Geostatistical Analyst. Figure 9.16 shows a grid generated using the 16 nearest neighbours with IDW.

Cross-validation was used to compare the three methods using 16 nearest neighbours. When using cross-validation to compare prediction approaches, it is usual to use summaries of the errors such as the mean error or the RMSE. As its name suggests, the RMSE is the root of the mean squared errors (i.e. take each error, square it, take the average of these squared errors, and then take the square root of this average); this is a widely used summary measure and is interpreted below. The mean error indicates bias: if it is negative then the procedure tends to under-predict values and when it is positive

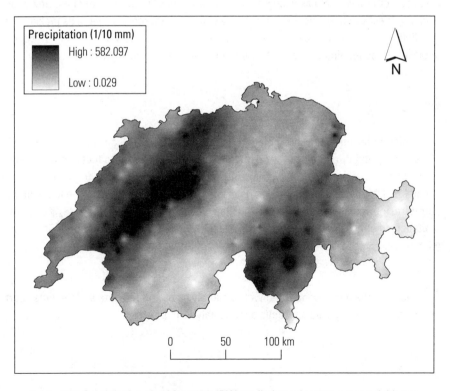

Figure 9.16 Precipitation on 8 May 1986: IDW prediction using 16 nearest neighbours.

it suggests over-prediction. Ideally, therefore, the mean error would be 0. The RMSE represents the magnitude of errors and a small RMSE value corresponds to accurate predictions. With IDW, the mean cross-validation error was −0.520 and the RMSE was 47.82. The figures for TPS were mean error −0.418 and RMSE 53.87. For OK the figures were mean error −0.189 and RMSE 47.81. In this case, the least biased predictions (mean error closest to 0) are provided by OK. The smallest errors overall (as measured by the RMSE) are also for OK, although the difference between the OK and IDW RMSE values is very small. Note that TPS predictions are the least accurate in this case, but that more accurate predictions are obtained when other variants of TPS (e.g. TPS with tension, which can be conducted in ArcGIS™, and is defined by Lloyd (2006)) are used.

Summary

This chapter provided an introduction to some of the most widely used approaches to the generation of surfaces from point data. In addition, a short outline of areal interpolation was given. In terms of selection of methods, it was noted that differences in prediction results are a function of sampling density and spatial variation. Approaches like cross-validation offer a way of assessing the performance of different approaches. However, such approaches should not be used blindly and other approaches, such as jackknifing (predicting to one set of locations (at which there are observed values) using a second data set and computing the errors of the predictions), are likely to be more robust.

Further reading

More information on spatial interpolation is provided by Burrough and McDonnell (1998), Lloyd (2006), and Chang (2008), for example. There are many introductions to geostatistics (e.g. Goovaerts, 1997; Armstrong, 1998; Webster and Oliver, 2007; Atkinson and Lloyd, 2009). Many different applications of interpolation procedures can be found in the literature. Interpolation has been used to map elevation (Lloyd and Atkinson, 2002), precipitation amount (Goovaerts 2000; Lloyd 2002, 2005, 2009, 2010), and airborne pollutants (Lloyd and Atkinson, 2004), amongst many other variables.

➡ The next chapter is concerned with the analysis of gridded data and the latter part of the chapter has a particular focus on the analysis of DEMs.

10

Analysis of grids and surfaces

10.1 Introduction

This chapter introduces methods for processing and analysing gridded data in general and gridded elevation data in particular. Topics include map algebra, image processing, and spatial filters. Derivatives of altitude, such as slope and other products derived from surfaces, are also described. Image processing has a central role in GIS contexts. This chapter begins by outlining some key approaches to the analysis of gridded data, of which images are a common example.

10.2 Map algebra

Most GIS software packages include functions for map algebra. In words, using map algebra raster layers can be combined in various ways. For example, the values in overlapping cells may be added together using this format:

OUTPUT = MAP1 + MAP2

Similarly, values could be multiplied or any other arithmetic operation applied. Obviously, in many cases such operations may be meaningless—it makes no sense to add altitude values to precipitation values. However, addition of pollutant scores (as opposed to concentrations), for example, may be sensible. Map algebra provides a means of masking files. For example, if a raster exists which has values of 1 for areas of interest and 0 for areas that are not of interest then this map layer could be multiplied

1	0	0		7	6	4		7	0	0
---	---	---		---	---	---		---	---	---
0	1	0		6	5	4		0	5	0
0	1	1		5	4	3		0	4	3

with $\times$ between the first two grids and $=$ between the second and third.

Figure 10.1 Retaining grid cells indicated by 1 in a binary grid.

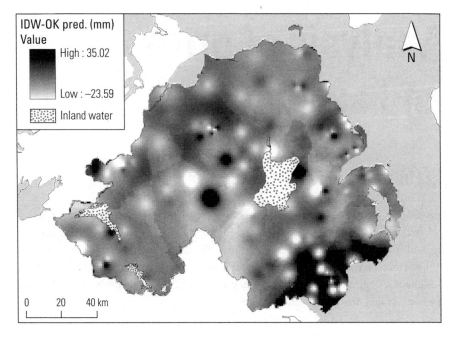

Figure 10.2 IDW minus OK predictions of precipitation in July 2006 (both using 16 nearest neighbours).

by any other layer. The result is that only features in areas of interest are transferred to the new raster layer as all cells of interest are multiplied by 1, whereas cells outside the area of interest are multiplied by 0. Figure 10.1 gives a simple example of this operation.

Figure 10.2 shows the map obtained when IDW predictions of precipitation (Figure 9.7) are subtracted from OK predictions (Figure 9.13). In this case, the map algebra command corresponds to OUTPUT = IDWMAP − OKMAP. Burrough and McDonnell (1998) provide more examples of map algebra.

Positive values indicate that the IDW predictions are larger while negative values indicate that OK predictions are larger. In other words, where values are positive, IDW predicts larger precipitation amounts than OK does and, where they are negative, OK predicts larger precipitation amounts than IDW does. This kind of approach offers a simple means to compare grids. One obvious potential applications area is the comparison of grids for different time periods. If digital elevation models (DEMs, as illustrated in Figure 10.5), for example, are available for two time periods, one could be subtracted from the other to assess rates of erosion. However, it is, of course, necessary to ensure that data sets compared in this way are compatible (e.g. they have the same spatial resolution and the data were collected with instruments which make measurements with a comparable level of accuracy). Matrix algebra is a key tool in multicriteria decision analysis (see Section 5.3).

10.3 Image processing

The analysis of images has, in comparison to GIS, a long history. The need to remove 'noise' from raster grids (or images; e.g. remotely sensed images) or to identify features contained in images has led to the development of a wide range of sophisticated tools. Tools for spatial segmentation and classification are widely used to deal in different ways with the grouping together of pixels that have, in some respect, similar characteristics. Such methods are outside the remit of this account. This section begins by defining different classes of grid operators. Grid operators can be divided into four groups (Chou, 1997):

- local functions: work on every single cell (a cell is treated as an individual object)
- focal functions: derive a new value based on the neighbourhood of a pixel
- zonal functions: work on each group of cells of identical values
- global functions: work on a cell based on the data in the entire grid.

The addition of two grids is a local function since the values in overlapping individual cells are combined. Moving window functions (such as spatial filters, described in the following section) are examples of focal operators. Zonal operators deal with values that fall within particular zones, and an example is given below. A common example of a global function is the computing of Euclidean (straight line) distance from one or more source locations to all cells in a grid (see Section 4.2).

With zonal operators, values from one input grid that fall within zones indicated by a second input grid are combined in some way. The zonal mean is illustrated in Figure 10.3. The zonal minimum or maximum, for example, could be computed instead. Note that the boundaries of the individual cells in the zones layer and the zonal means layer are not shown—the values in both cases would be written to all cell locations indicated in the values grid. As an example, for zone 1: $(45+44+44+43+42+43+42)/7=43.286$.

There are many potential applications areas for zonal operations, for example average erosion in construction zones. Any application which makes use of zones represented as rasters might make use of operators like those just described.

The focus of the following sections is spatial filters for image smoothing or enhancing edges of features.

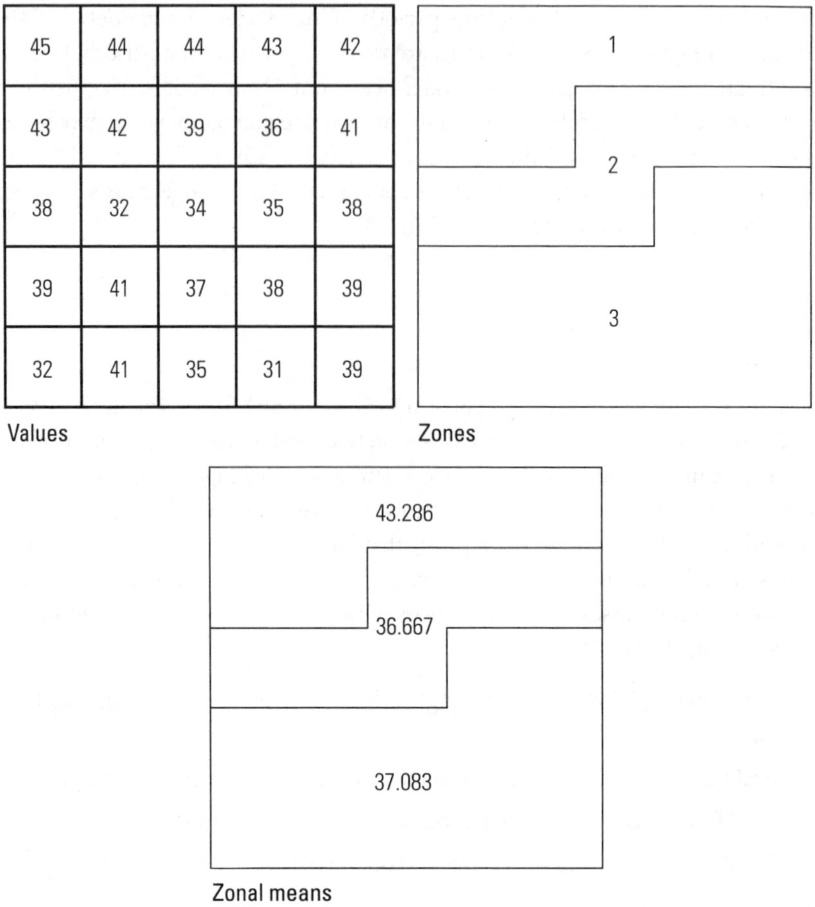

45	44	44	43	42
43	42	39	36	41
38	32	34	35	38
39	41	37	38	39
32	41	35	31	39

Values

Zones

1

2

3

Zonal means

43.286

36.667

37.083

Figure 10.3 Zonal mean.

10.4 Spatial filters

The idea of moving windows for processing images was introduced in Section 4.6. Such operations are called spatial filters. The application of a mean filter, as illustrated in Section 4.6, results in a smoothed version of the original image. In contrast, application of a standard deviation filter will result in a new image that highlights the edges of features in the original image. Figure 10.4 shows an input grid (the same one as used in Section 4.6) and the output of a mean filter and a standard deviation filter.

Comparison of the values of the standard deviation output in Figure 10.4 with the input grid shows how areas with more variable values have larger standard deviations. As with the example in Section 4.6, the edge pixels are not included in the outputs.

Figure 10.5 shows elevations in Northern Ireland. Figure 10.6 shows the mean of elevation for a 3×3 pixel moving window. Comparison with the original elevation map shows that the range of values in the mean map is smaller than in the original map as the effect of outliers (values that are large or small locally) is reduced through

Output grid
(3 × 3 cells)
location
is shaded

45	44	44	43	42
43	42	39	36	41
38	32	34	35	38
39	41	37	38	39
32	41	35	31	39

Input grid

40.11	38.78	39.11
38.33	37.11	37.44
36.56	36.00	36.22

Mean

4.68	4.66	3.62
3.61	3.26	2.19
3.50	3.57	2.68

Standard deviation

Figure 10.4 Mean average and standard deviation computed for a 3×3 pixel moving window.

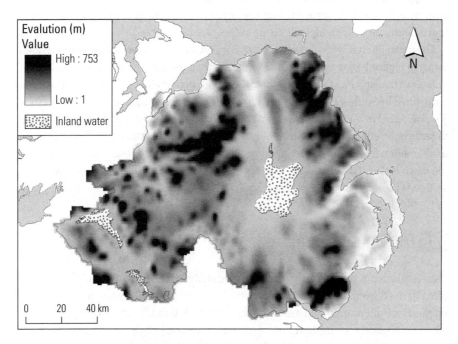

Figure 10.5 Elevation: Northern Ireland. Data available from the United States Geological Survey/EROS, Sioux Falls, SD.

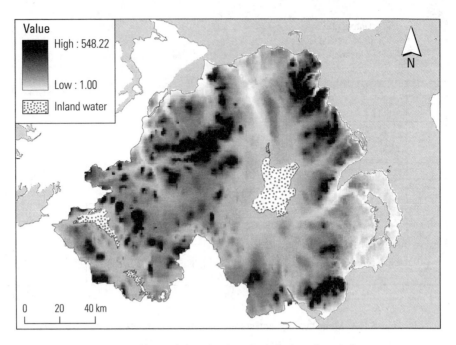

Figure 10.6 Mean of elevation for a 3×3 pixel moving window.

averaging. Figure 10.7 shows the standard deviation, also for a 3×3 pixel moving window. In this case, it is apparent that the edges of more notable topographic features (particularly those with high elevation values) are highlighted.

The mean filtered DEM (recall that this is a grid-based representation of topographic form) in Figure 10.6 is smoother in appearance than the unfiltered DEM in Figure 10.5. This is particularly clear on the edges of areas with higher elevation values. Other filters are defined by Sonka *et al.* (1999) and Lloyd (2006). Smoothing filters (such as the mean filter) are used in many different contexts. For example, smoothing filters are often used to reduce the effect of 'noise' in remotely sensed images (Mather, 2004), although the median filter, for example, may be more suitable in such cases than the mean filter (as the mean is affected by outliers, whereas the median is not). The focus of the following section is the analysis of DEMs specifically (as opposed to other forms of raster grids).

10.5 Derivatives of altitude

The remainder of this chapter focuses on the analysis of DEMs. Various products are often derived from DEMs—derivatives such as gradient and aspect (the two making up slope) are often computed. Gradient refers to the maximum rate of change in altitude while aspect refers to the direction of the maximum rate of change (e.g. gradient may be north facing) (Burrough and McDonnell, 1998). The terms 'gradient' and 'slope' are sometimes used interchangeably, but here the convention of defining slope to comprise

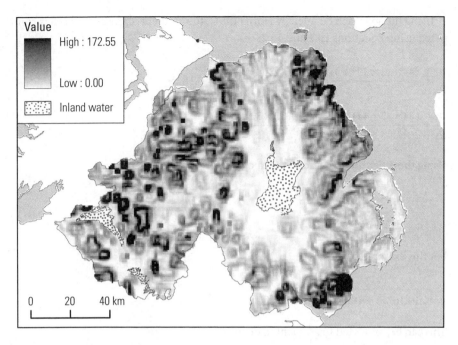

Figure 10.7 Standard deviation of elevation for a 3×3 pixel moving window.

gradient and aspect together is preferred. There are several methods for deriving gradient and one of the most popular is demonstrated here. Given cells labelled as follows:

$$z_1 \quad z_2 \quad z_3$$
$$z_4 \quad z_5 \quad z_6$$
$$z_7 \quad z_8 \quad z_9$$

the gradient can be estimated from (Horn, 1981):

$$\frac{\partial z}{\partial x} = \frac{(z_3 + (2z_6) + z_9) - (z_1 + (2z_4) + z_7)}{8h_x} \tag{10.1}$$

$$\frac{\partial z}{\partial y} = \frac{(z_3 + (2z_2) + z_1) - (z_9 + (2z_8) + z_7)}{8h_x} \tag{10.2}$$

where h_x is the grid spacing in the x direction (east–west) and h_y is the grid spacing in the y direction (north–south). The method is illustrated using the following 3×3 cell matrix, which we will assume has a spatial resolution of 10 m (i.e. cells that cover an area in the real world of 10×10 m):

45   44   44
43   42   39
38   32   34

Using the approach detailed by Horn (1981), the gradient of the central cell is computed using Equations 10.1 and 10.2:

$$\frac{\partial z}{\partial x} = \frac{(44 + (2 \times 39) + 34) - (45 + (2 \times 43) + 38)}{(8 \times 10)} = 0.1625$$

$$\frac{\partial z}{\partial y} = \frac{(44 + (2 \times 44) + 45) - (34 + (2 \times 32) + 38)}{(8 \times 10)} = 0.5125$$

Using these figures, the gradient is calculated from:

$$g = \sqrt{\left(\frac{\partial z}{\partial x}\right)^2 + \left(\frac{\partial z}{\partial y}\right)^2} \tag{10.3}$$

For this example, this leads to:

$$g = \sqrt{(0.1625)^2 + (0.5125)^2} = 0.5376 \text{ m}$$

This can be converted to gradient in degrees (gd):

$$gd = \text{atan}(g) \times 57.29578 \tag{10.4}$$

where atan (also given by $\tan^{-1}$) is the inverse tangent (see Appendix C). Gradient is often also expressed as percentages. Here gradient in degrees is:

$$gd = 0.4933 \times 57.29578 = 28.264°$$

Figure 10.8 shows gradient (in degrees) derived from a DEM of Northern Ireland. Note that the spatial resolution of the DEM from which gradient was derived is 740.224 m. The gradients are never as large as in the computed example above because any dramatic gradients are, in effect, averaged out because of the coarse spatial resolution.

Note that the gradient map is visually similar to the standard deviation map given earlier in the chapter (Figure 10.7). This is not surprising as the standard deviation picks out the edges of features and gradient does the same. Maps of gradient are used widely for modelling erosion. Mitásová *et al.* (1996) computed gradient and aspect as a part of a set of procedures for modelling erosion and deposition. Li *et al.* (2004) provide a detailed account of gradient and aspect derivation as well as other terrain parameters.

10.6 Other products derived from surfaces

Besides gradient and other derivatives of altitude, many other products are derived from DEMs. These include maps indicating the direction of steepest downhill descent

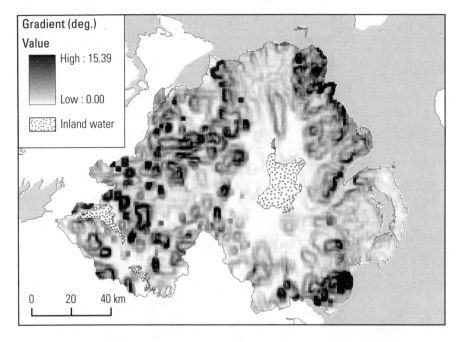

Figure 10.8 Gradient in degrees.

(routing), irradiance, watershed, cost surfaces, and visibility maps. Algorithms also exist for processing DEMs—in particular, removal of 'anomalous' pits or peaks is often a concern. This section considers two classes of widely used DEM-based approaches: visibility analysis and derivation of cost surfaces.

Visibility analysis is concerned with the identification of areas in the landscape that are visible from a given location (or a set of locations). Many GIS offer tools for generating viewsheds—areas that are visible from a given point. Figure 10.9 indicates a viewing cell in black and the pixel that is to be tested for visibility is grey. Figure 10.10 indicates the two possible cases: that the cell is invisible or visible. By testing the line of sight to each cell, a map of visible cells (the viewshed) can be generated.

The line of sight from one location to another can be assessed using:

$$h_{crit} = \frac{d_{vo}h_t + d_{to}h_v}{d_{to} + d_{vo}}$$

(10.5)

where h_{crit} is the critical value for the height of an obstacle, d_{to} is the distance between the target and the obstacle, d_{vo} is the distance between the viewer and the obstacle, h_t is the height of the target, and h_v is the height of the viewer. If the height of the obstacle is less than h_{crit} then the target is visible (Li *et al.*, 2004). As an example, for $d_{to}=24.5$, $d_{vo}=12$, $h_t=8$, and $h_v=5$:

$$h_{crit} = \frac{12 \times 8 + 24.5 \times 5}{24.5 + 12} = \frac{218.5}{36.5} = 5.986$$

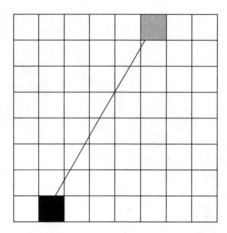

Figure 10.9 Viewing direction.

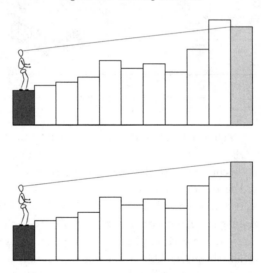

Figure 10.10 Top: invisible cell; bottom: visible cell (both shaded grey).

In this case, if the obstacle is higher than 5.986 units then the target will not be visible to the viewer.

When the height of the obstacle (h_o) is known, the length of the possible line of sight that is blocked by the obstacle can be computed using (Li *et al.*, 2004):

$$d_{to} = \frac{h_o}{h_v + h_o} \times d_{vo} \tag{10.6}$$

This approach could be applied to all cells in a raster image, but more efficient algorithms exist. Figure 10.11 shows a DEM with a viewpoint superimposed (A) and the viewshed from this location (B).

Note that the viewer height must be specified and not simply the elevation or building height at the viewing location. Of course, visibility can only be meaningfully

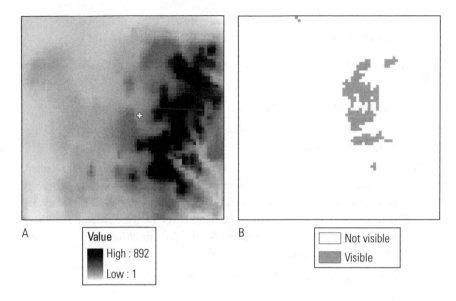

Figure 10.11 DEM (with elevations in metres) with a viewpoint superimposed (A) and the viewshed from this location (B).

measured up to the distance to which objects will be visible and such considerations must be taken into account when generating viewsheds. Viewsheds are used widely in various contexts. One obvious use is generating viewsheds as part of the planning process whereby the visual impact of a development on the landscape can be minimized by identifying areas that are not visible from particular locations. Wheatley (1995) used viewshed analysis to explore intervisibility between Neolithic long barrows in the Danebury region (in south-west England). Fisher (1992) considers the uncertainties inherent in viewsheds.

Another frequently generated product using DEMs is a cost surface. Cost surfaces indicate the minimum cost of reaching a cell from source cells. To generate a cost surface a friction surface and source cells are needed. The friction surface is often simply a gradient map—steep gradients represent a greater degree of friction or higher cost. A cost surface can be generated using Dijkstra's algorithm, as outlined in Section 6.5. When a DEM is generated using spatial interpolation, the interpolation method used will determine how 'rough' the surface is and thus the modelled 'cost' of travelling over the surface. Such factors, which may not relate to variations in the real world, should be taken into account.

An example is given using the cost grid (e.g. gradient) and source cells (cell locations from which distance are measured) shown in Figure 10.12.

Cost distances are computed as follows:

- neighbours sharing an edge = average of costs in the neighbouring cells
  (e.g. $45 + 44 = {}^{89}/2 = 44.5$)

- neighbours sharing only a corner=average of costs in the neighbouring cells multiplied by 1.4142 (e.g. $(45+42)/2 = 87/2 = 43.5 \times 1.1412 = 61.5$).

The cost surface procedure (note the connections with Section 6.5) can be outlined as follows:

- Compute the cost distances from each source cell to its neighbours.

- Select the smallest cost distance and compute the smallest cost distance to that cell's neighbours—these cells are activated.

- The next activated cell with the smallest accumulative cost distance is selected. Next, compute the smallest cost distance to that cell's neighbours. Every time a cell becomes accessible to a source cell through a different path it is reactivated and its accumulative cost must be recalculated because the new path may have a smaller accumulative cost (Chang, 2008). If it does not, then the accumulative cost value remains the same.

- Continue this process until all smallest accumulative cost distances have been computed.

Cost grid:

45	44	44	43
43	42	39	36
38	32	34	35
39	41	37	38

Cost grid

Source cells:

1			
			2

Source cells

Figure 10.12 Cost grid and source cells.

Step 1:

0	44.5		
44	61.5		
		50.9	36.5
		37.5	0

Step 1

Step 2:

0	44.5		
44	61.5	87.4	72
		50.9	36.5
		37.5	0

Step 2

Step 3:

0	44.5		
44	61.5	87.4	72
	83.9	50.9	36.5
	76.5	37.5	0

Step 3

Step 4:

0	44.5		
44	61.5	87.4	72
84.5	83.9	50.9	36.5
	76.5	37.5	0

Step 4

Step 5:

0	44.5	88.5	
44	61.5	87.4	72
84.5	83.9	50.9	36.5
	76.5	37.5	0

Step 5

Step 6:

0	44.5	88.5	111.5
44	61.5	87.4	72
84.5	83.9	50.9	36.5
	76.5	37.5	0

Step 6

Step 7:

0	44.5	88.5	111.5
44	61.5	87.4	72
84.5	83.9	50.9	36.5
116.5	76.5	37.5	0

Step 7: Least accumulative cost distances

Figure 10.13 Cost surface derivation.

The procedure is illustrated in Figure 10.13. The smallest accumulative cost distances that have not yet been included are indicated by bold characters.

In step 1, the cost distances are measured from the source cells, and the smallest cost distance (36.5) is selected. The cost distances are then measured from the selected cell and the smallest non-selected cost distance (37.5) is selected, the end result being step 2. Next, the cost distances are measured from that cell location and the smallest non-selected value is selected (44)—this is step 3. The cost distances are then measured from the selected cell and the smallest non-selected cost distance is selected (44.5), leading to step 4. The cost distances are measured from this cell and the smallest non-selected cost distance (72) is selected—this is step 5. The cost distances are measured from this cell location and the smallest non-selected value (76.5) is selected. After the distances are measured from that cell all accumulative cost distances have been computed and the final grid is step 7.

The direction of the least accumulative cost path for the above example is given in Figure 10.14 while the origin of the least cost path (i.e. the parent cell) is given in Figure 10.15 (1 and 2 being the values assigned to the source cells).

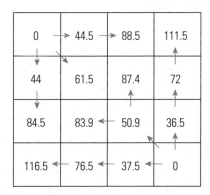

Figure 10.14 Direction of the least accumulative cost path.

1	1	1	2
1	1	2	2
1	2	2	2
2	2	2	2

Figure 10.15 Origin of the least accumulative cost path.

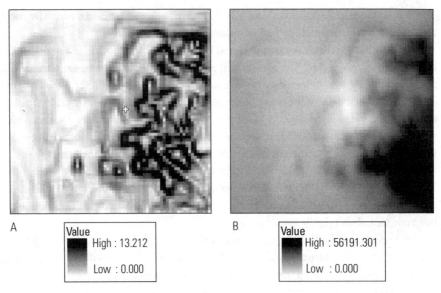

Figure 10.16 Gradient (in degrees) with starting location superimposed (A) and the cost surface from this location (B).

Figure 10.16 shows a gradient map (in degrees) with the starting location (A) and the cost surface derived using these inputs (B).

Cost surfaces are computed in numerous different contexts. In an application concerned with identifying possible routes for the construction of a new railway, Cowen *et al.* (2000) generated cost surfaces accounting for several different factors, including grade (difference between existing and desired elevation), road crossings, stream crossings, and track cost. The generation of alternative routes and comparisons between them has been found to be very useful in many planning projects of this kind.

10.7 Case study

Gradient (in degrees) was computed using ArcGIS™ Spatial Analyst from the DEM (with a spatial resolution of 1009.975 m) illustrated in Figure 10.17. The DEM is described by Dubois (2003). The algorithm used is that detailed by Horn (1981) and the output is shown in Figure 10.18.

The southern part of Switzerland is dominated by the Alps, and this explains the steep gradients in that region; in the north-west the Jura mountains are evident. In terms of, for example, modelling the costs of transporting goods, this output would suggest that it requires more effort to move goods over land in the south of Switzerland than in the north. If a value can be assigned to a gradient derived from a DEM then such data may provide the basis of a useful analysis.

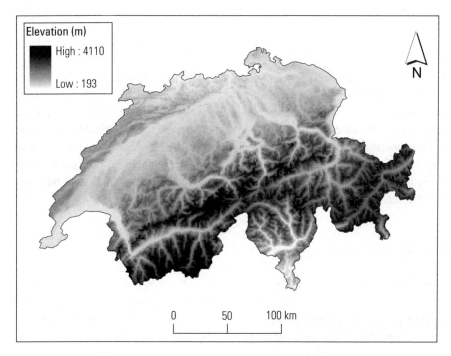

Figure 10.17 Elevation: Switzerland.

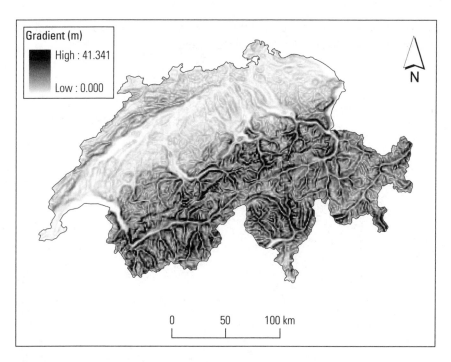

Figure 10.18 Gradient: Switzerland.

Summary

This chapter has dealt with various aspects of surface and analysis. Map algebra, image processing, and spatial filters were introduced. Following these, derivation of gradient was detailed before a summary of some widely used methods for the analysis of surfaces was given. Such approaches are very widely used in many different applications and an understanding of such approaches is essential in developing a rounded knowledge of spatial data analysis. As detailed in Section 2.8.3, remote sensing is a core source of spatial data and processing the imagery, using the kinds of techniques introduced in this chapter, is an important task if these data are to be used to the full.

Further reading

Some basic principles of image processing are outlined by Burrough and McDonnell (1998), who also discuss elevation derivatives. The book by Sonka *et al.* (1999) provides a detailed introduction to some key principles of image processing. In remote sensing data contexts specifically, the book by Mather (2004) gives an excellent summary of some important issues. A dedicated account of DEMs and their analysis is given by Li *et al.* (2004).

➡ The next chapter revisits some of the themes explored throughout this book and suggests some ways forward for the interested reader.

Summary

11.1 Review of key concepts

The book has introduced a range of general concepts and specific approaches for spatial data analysis. From data models to aspatial statistics, moving windows, geographical weights, and spatial autocorrelation to methods for overlay of vector features, local regression, analysis of point patterns, and spatial interpolation, the focus has been broad. Chapter 2 introduced a range of core ideas, including data models, databases, spatial scale, spatial data collection, data errors, visualization, and simple data queries. In Chapter 3, some basic statistical approaches were detailed to give background for the spatial statistical methods presented later in the book. Chapter 4 was concerned initially with measurement of distances and areas. The following sections dealt with moving windows, geographical weights, the concepts of spatial dependence and autocorrelation, and the ecological fallacy, and the modifiable areal unit problem. In Chapter 5, overlay of different features and identification of areas fulfilling a range of criteria were the focus. Chapter 6 dealt with the characterization and analysis of networks. Chapter 7 outlined some approaches for analysing point patterns. In Chapter 8, various methods for exploring spatial patterning in variables were detailed. In particular, approaches for characterizing spatial structure in single variables (i.e. the degree to which values close together in space are similar) and methods for analysing geographical variations in the relationships between different variables were illustrated. Chapter 9 presented methods for spatial interpolation and Chapter 10 was concerned with the analysis of grids and surfaces.

While the focus is quite broad there are links running through the book that relate to the general approaches used. For example, the methods outlined in Chapters 7, 8, 9 and 10 make use of geographical weighting schemes. Spatial data analysis encompasses a very large array of methodological frameworks developed to overcome a huge variety of different problems. The range of approaches introduced has been from methods simply for exploring spatial patterns, to methods for assessing how meaningful (in some sense) those patterns might be, through to methods for finding optimal solutions to problems (e.g. the shortest path or the most suitable area) and methods for estimating data values at locations where there is no sample. This book could have been organized in many different ways and the balance of coverage of different methods could have been altered markedly. Nonetheless, it is hoped that it covers in reasonable detail a sufficient variety of kinds of approaches, if not specific methods, to demonstrate clearly the diversity of approaches.

There is very much more that could be said about the issues raised in this book and some suggestions as to how to proceed and further develop knowledge of spatial data analysis are given in Section 11.4. Appendix G provides a list of some problems and the corresponding solutions detailed in this book.

11.2 Other issues

There are many issues touched on within this book that could have been developed substantially, and references to additional sources are given to allow readers to expand their knowledge. There are also many issues that could have been discussed, but which were considered outside the remit of the book. However, this book is intended as a starting point that provides pointers to other sources where necessary and the coverage of material within the book is necessarily focused on particular issues.

Many specific issues that some readers might like to have seen covered are omitted, and this is necessarily the case because of limitations of space and because the key aim of this book is to introduce in a focused way a relatively limited array of key concepts and methods. As an example, the book discusses connectivity in terms of identifying neighbouring areas. An important area of research, with many applications, concerns connectivity, as well as the form of landscape areas. McGarigal and Marks (1995) present a software package, Fragstats, which is designed to quantify landscape structure (e.g. by size or shape of landscape patches or their density over an area). Such approaches are not yet a standard part of GIS software and are, therefore, not discussed in this book. A more general omission is geographic data mining. The theme of geographic data mining (Shekar *et al.*, 2003; Miller, 2008) is an important one. Geographic knowledge discovery, with, as a core component, geographic data mining is based on the belief that there is new and useful knowledge to be extracted from the vast array of geographic data sources now available (Miller, 2008). The development and use of dynamic models is central to many applications, but this is another topic excluded from this book. It is hoped that the sources cited will provide the necessary information where some topics are covered only briefly in this book or omitted completely.

11.3 Problems

This book presents a variety of solutions to problems, but it only makes passing reference to other core issues that have an impact on spatial data and their analysis. There are widely used alternatives to most of the methods detailed this book and the selection of appropriate methods is often not straightforward. Since there are many alternative approaches for the spatial analyst to choose from, often experience is important. As an example, there are numerous methods for spatial interpolation (the focus of Chapter 9) and choice of specific method, interpolation neighbourhood (use all data or a local subset), and use of exact or approximate interpolation, are all likely to be important. Guidance has been offered in the text, although this can be no substitute for experimentation and direct assessment of the differences in results obtained using different combinations of approaches. Descriptions of other methods are provided in the publications referenced in Section 11.4.

Expensive technology and sophisticated-looking software cannot disguise the fact that spatial data analysis is based on models that may be poor abstractions of reality. These limitations and the various kinds of errors that affect any data source must be considered. The propagation of errors from one stage of processing to another (as referred to in Section 2.9.1) is the subject of much research (see Burrough and McDonnell, 1998) and all users of (spatial) data are obliged to consider the quality of their data and possible impacts on analyses that are based on these data.

11.4 Where next?

GIS and spatial data analysis are practical topics. If the tools offered are not used then there is no point in them. To begin to develop an understanding of how and why particular approaches are employed and to become aware of their shortcomings, it is necessary to apply the methods. In short, experimentation is a vital part of the learning process. The book website provides details of some software packages that implement the methods described in the book. Commercial packages like ArcGIS™ include all of the basic analytical functions that most users are likely to need. Indeed almost all of the case studies in this book can be directly replicated using ArcGIS™ and its associated extensions. Some very extensive and powerful packages are completely free and there is, therefore, not necessarily any financial barrier to sophisticated spatial data analysis. For example, the GIS GRASS (Neteler and Mitásová, 2007) and the R programming environment[1] (Bivand *et al.*, 2008) and associated routines offer sophisticated functionality at no cost.

In terms of reading material, there are several book chapters and full books that may make sensible next steps. Brief summaries of methods for the analysis of spatial data which provide pointers to more extensive material are provided by Anselin (2005), Getis (2005), Fischer (2005), and Charlton (2008). The book by O'Sullivan and Unwin (2002) expands on some of the issues discussed here. Bailey and Gatrell (1995), Lloyd (2006), and De Smith *et al.* (2007) provide further accounts of methods discussed in this book. Many books have been written for particular audiences. A good example is the book by Plane and Rogerson (1994), which deals with the analysis of data about human

---

1 http://www.r-project.org/

populations. For more specific material, see the further reading sections at the end of each chapter.

There are many other issues that are the focus of research efforts. Looking at key journals such as the *International Journal of Geographical Information Science*, *Transactions in GIS*, *International Journal of Remote Sensing*, *Computers and Geosciences*, and others will reveal the range and extent of research that is ongoing. As with any well-developed discipline, it is easy to become lost in the detail of a plethora of diverse applications. However, there are plenty of introductory starting points to guide your efforts, as the further reading lists in this book suggest. The book website provides specific material to support use of this book as well as links to other relevant websites and information about other material that may be useful.

Summary and conclusions

This book has merely scraped the surface of spatial data analysis. Hopefully, it will have successfully introduced some new ideas that will aid understanding of other issues discussed in the literature. It is also hoped that it has assisted more in developing understanding than in generating confusion. The analysis of spatial data is a commercially and academically highly significant field. It is hoped that this book will have introduced some approaches and concepts to a broader readership and illustrated something of the breadth of tools available and the problems that they can be used to solve.

References

Aitchison, J. (1986) *The Statistical Analysis of Compositional Data*. London: Chapman and Hall.

Anselin, L. (1995) Local indicators of spatial association—LISA. *Geographical Analysis*, 27, 93–115.

Anselin, L. (1996) The Moran scatterplot as an ESDA tool to assess local instability in spatial association. In Fischer, M. M., Scholten, H., and Unwin, D. (eds), *Spatial Analytical Perspectives on GIS*. London: Taylor and Francis, pp. 111–125.

Anselin, L. (2005) Interactive techniques and exploratory data analysis. In Longley, P. A., Goodchild, M. F., Maguire, D. J., and Rhind, D. W. (eds), *Geographical Information Systems: Principles, Techniques, Management and Applications*. 2nd edn, abridged. Hoboken, NJ: John Wiley & Sons Ltd, pp. 253–266.

Anselin, L., Syabri, I., and Kho, Y. (2006) GeoDa: An introduction to spatial data analysis. *Geographical Analysis*, 38, 5–22.

Armstrong, M. (1998) *Basic Linear Geostatistics*. Berlin: Springer-Verlag.

Atkinson, P. M. and Lloyd, C. D. (2009) Geostatistics and spatial interpolation. In Fotheringham, A. S. and Rogerson, P. A. (eds), *The SAGE Handbook of Spatial Analysis*. London: SAGE Publications, 159–181.

Atkinson, P. M. and Tate, N. J. (2001) Five key recommendations for GI Science. In Tate, N. J. and Atkinson, P. M. (eds), *Modelling Scale in Geographical Information Science*. Chichester: John Wiley & Sons Ltd, pp. 263–271.

Bailey, T. C. and Gatrell, A. C. (1995) *Interactive Spatial Data Analysis*. Harlow: Longman Scientific and Technical.

Bannister, A., Raymond, S., and Baker, R. (1998) *Surveying*. 7th edn. Harlow: Pearson Education.

Birkin, M., Clarke, G., Clarke, M., and Wilson, A. (1996) *Intelligent GIS: Location Decisions and Strategic Planning*. Cambridge: GeoInformation International.

Bivand, R. S., Pebesma, E. J., and Gómez-Rubio, V. (2008) *Applied Spatial Data Analysis with R*. New York: Springer.

Bonham-Carter, G. F. (1994) *Geographic Information Systems for Geoscientists*. Kidlington: Pergamon.

Brown, J. D. and Heuvelink, G. B. M. (2008) On the identification of uncertainties in spatial data and their quantification with probability distribution functions. In Wilson, J. P. and Fotheringham, A. S. (eds), *The Handbook of Geographic Information Science*. Maldon, MA: Blackwell Publishing, pp. 94–107.

Brunsdon, C. (1995) Estimating probability surfaces for geographical point data: an adaptive algorithm. *Computers and Geosciences*, 21, 877–894.

Brunsdon, C. (2008) Inference and spatial data. In Wilson, J. P. and Fotheringham, A. S. (eds), *The Handbook of Geographic Information Science*. Maldon, MA: Blackwell Publishing, pp. 337–351.

Brunsdon, C., McClatchey, J., and Unwin, D. J. (2001) Spatial variations in the average rainfall-altitude relationship in Great Britain: an approach using geographically weighted regression. *International Journal of Climatology*, 21, 455–466.

Burrough, P. A. and McDonnell, R. A. (1998) *Principles of Geographical Information Systems*. Oxford: Oxford University Press.

Chang, K. (2008) *Introduction to Geographic Information Systems*. 4th edn. Boston: McGraw-Hill.

Charlton, M. E. (2008) Quantitative methods and geographic information systems. In Wilson, J. P. and Fotheringham, A. S. (eds), *The Handbook of Geographic Information Science*. Maldon, MA: Blackwell Publishing, pp. 379–394.

Chou, Y.-H. (1997) *Exploring Spatial Analysis in Geographic Information Systems*. Albany: OnWord Press.

Clarke, K. (1999) *Getting Started with Geographic Information Systems*. 2nd edn. Upper Saddle River, NJ: Prentice Hall.

Cliff, A. D. and Ord, J. K. (1973) *Spatial Autocorrelation*. London: Pion.

Codd, E. F. (1970) A relational model of data for large shared data banks. *Communications of the Association for Computing Machinery*, 13, 377–387.

Conolly, J. and Lake, M. (2006) *Geographical Information Systems for Archaeologists*. Cambridge: Cambridge University Press.

Cowen, D. J., Jensen, J. R., Hendrix, C., Hodgson, M. E., and Schill, S. R. (2000) A GIS-assisted rail construction econometric model that incorporates LiDAR data. *Photogrammetric Engineering and Remote Sensing*, 66, 1323–1328.

Cressie, N. A. C. (1993) *Statistics for Spatial Data*. Revised edition. New York: John Wiley & Sons Ltd.

D

Delmelle, E. (2009) Spatial sampling. In Fotheringham, A. S. and Rogerson, P. A. (eds), *The SAGE Handbook of Spatial Analysis*. London: SAGE Publications, pp. 183–206.

De Smith, M. J., Goodchild, M., and Longley, P. A. (2007) *Geospatial Analysis. A Comprehensive Guide to Principles, Techniques and Software Tools*. 2nd edn. Leicester: Matador.

Diggle, P. J. (2003) *Statistical Analysis of Spatial Point Patterns*. 2nd edn. London: Arnold.

Dijkstra, E. W. (1959) A note on two problems in connexion with graphs. *Numerische Mathematik*, 1, 269–271.

Dubois, G. (2003) Spatial interpolation comparison 97. Introduction and description of the data set. In Dubois, G., Malczewski, J., and De Cort, M. (eds), *Mapping Radioactivity in the Environment. Spatial Interpolation Comparison 97*. Luxembourg: Office for Official Publications of the European Communities, pp. 39–44.

E

Ebdon, D. (1985) *Statistics in Geography*. 2nd edn. Oxford: Blackwell.

F

Fischer, M. M. (2005) Spatial analysis: retrospect and prospect. In Longley, P. A., Goodchild, M. F., Maguire, D. J., and Rhind, D. W. (eds), *Geographical Information Systems: Principles, Techniques, Management and Applications*. 2nd edn, abridged. Hoboken, NJ: John Wiley & Sons Ltd, pp. 283–292.

Fisher, P. F. (1992) First experiments in viewshed uncertainty: simulating the fuzzy viewshed. *Photogrammetric Engineering and Remote Sensing*, 58, 345–352.

Fotheringham, A. S., Brunsdon, C., and Charlton, M. (2000) *Quantitative Geography: Perspectives on Spatial Data Analysis*. London: SAGE Publications.

Fotheringham, A. S., Brunsdon, C., and Charlton, M. (2002) *Geographically Weighted Regression: The Analysis of Spatially Varying Relationships*. Chichester: John Wiley & Sons Ltd.

G

Getis, A. (2005) Spatial statistics. In Longley, P. A., Goodchild, M. F., Maguire, D. J., and Rhind, D. W. (eds), *Geographical Information Systems: Principles, Techniques, Management and Applications*. 2nd edn, abridged. Hoboken, NJ: John Wiley & Sons Ltd, pp. 239–251.

Getis, A. and Ord, J. K. (1996) Local spatial statistics: an overview. In Longley, P. and Batty, M. (eds), *Spatial Analysis: Modelling in a GIS Environment*. Cambridge: Geo-Information International, pp. 261–277.

Goovaerts, P. (1997) *Geostatistics for Natural Resources Evaluation*. New York: Oxford University Press.

Goovaerts, P. (2000) Geostatistical approaches for incorporating elevation into the spatial interpolation of rainfall. *Journal of Hydrology*, 228, 113–129.

Gregory, I. N. and Ell, P. S. (2007) *Historical GIS: Technologies, Methodologies and Scholarship*. Cambridge: Cambridge University Press.

Gregory, S. (1968) *Statistical Methods and the Geographer*. 2nd edn. London: Longman.

Griffith, D. A. (1988) *Advanced Spatial Statistics: Special Topics in the Exploration of Quantitative Spatial Data Series*. Dordrecht: Kluwer.

H

Haining, R. (2003) *Spatial Data Analysis: Theory and Practice*. Cambridge: Cambridge University Press.

Heywood, I., Cornelius, S., and Carver, S. (2006) *An Introduction to Geographical Information Systems*. 3rd edn. Harlow: Pearson Education.

Hill, E. G., Ding, L., and Waller, L. A. (2000) A comparison of three tests to detect general clustering of a rare disease in Santa Clara County, California. *Statistics in Medicine*, 19, 1363–1378.

Horn, B. K. P. (1981) Hill shading and the reflectance map. *Proceedings of the Institute of Electrical and Electronics Engineers*, 69, 14–47.

Hudson, G. and Wackernagel, H. (1994) Mapping temperature using kriging with external drift: theory and an example from Scotland. *International Journal of Climatology*, 14, 77–91.

Hutchinson, M. F. (1989) A new procedure for gridding elevation and stream line data with automatic removal of spurious pits. *Journal of Hydrology*, 106, 211–232.

Isaaks, E. H. and Srivastava, R. M. (1989) *An Introduction to Applied Geostatistics*. New York: Oxford University Press.

Journel, A. G. and Huijbregts, C. J. (1978) *Mining Geostatistics*. London: Academic Press.

Kitchin, R. and Tate, N. (2000) *Conducting Research into Human Geography: Theory, Methodology and Practice*. Harlow: Prentice Hall.

Laurini, R. and Thompson, D. (1992) *Fundamentals of Spatial Information Systems*. APIC Series 37. London: Academic Press.

Lee, J. and Wong, D. (2000) *Statistical Analysis with ArcView GIS*. Hoboken, NJ: John Wiley & Sons Ltd.

Li, F. S. and Zhang, L. J. (2007) Comparison of point pattern analysis methods for classifying the spatial distributions of spruce-fir stands in the north-east USA. *Forestry*, 80, 337–349.

Li, Z., Zhu, Q., and Gold, C. (2004) *Digital Terrain Modeling: Principles and Methodology*. Boca Raton, FL: CRC Press.

Lillesand, T. M., Kiefer, R. W., and Chipman, J. W. (2007) *Remote Sensing and Image Interpretation*. 6th edn. Hoboken, NJ: John Wiley & Sons Ltd.

Lilley, K. D., Lloyd, C. D., and Trick, S. (2007) Designs and designers of medieval 'new towns' in Wales. *Antiquity*, 81, 279–293.

Lloyd, C. D. (2002) Increasing the accuracy of predictions of monthly precipitation in Great Britain using kriging with an external drift. In Foody, G. M. and Atkinson, P. M. (eds), *Uncertainty in Remote Sensing and GIS*. Chichester: John Wiley & Sons Ltd, pp. 243–267.

Lloyd, C. D. (2004) Landform and Earth surface. In Atkinson, P. M. (ed.), *Geoinformatics: Encyclopedia of Life Support Systems*. Developed under the auspices of the UNESCO. Oxford: EOLSS Publishers. http://www.eolss.net.

Lloyd, C. D. (2005) Assessing the effect of integrating elevation data into the estimation of monthly precipitation in Great Britain. *Journal of Hydrology*, 308, 128–150.

Lloyd, C. D. (2006) *Local Models for Spatial Analysis*. Boca Raton, FL: CRC Press.

Lloyd, C. D. (2009) Multivariate interpolation of monthly precipitation amount in the United Kingdom. In Atkinson, P. M. and Lloyd, C. D. (eds), *Geostatistics for Environmental Applications: Proceedings of the Seventh European Conference on Geostatistics for Environmental Applications*. Berlin: Springer, in press.

Lloyd, C. D. (2010) Nonstationary models for exploring and mapping monthly precipitation in the United Kingdom. *International Journal of Climatology*, in press.

Lloyd, C. D. and Atkinson, P. M. (2002) Deriving DSMs from LiDAR data with kriging. *International Journal of Remote Sensing*, 23, 2519–2524.

Lloyd, C. D. and Atkinson, P. M. (2004) Increased accuracy of geostatistical prediction of nitrogen dioxide in the United Kingdom with secondary data. *International Journal of Applied Earth Observation and Geoinformation*, 5, 293–305.

Lloyd, C. D. and Lilley, K. D. (2009) Cartographic veracity in medieval mapping: analyzing geographical variation in the Gough Map of Great Britain. *Annals of the Association of American Geographers*, 99, 27–48.

Lloyd, C. D. and Shuttleworth, I. G. (2005) Analysing commuting using local regression techniques: scale, sensitivity and geographical patterning. *Environment and Planning A*, 37, 81–103.

Longley, P. A., Goodchild, M. F., Maguire, D. J., and Rhind, D. W. (2005a) *Geographic Information Systems and Science*. 2nd edn. Chichester: John Wiley & Sons Ltd.

Longley, P. A., Goodchild, M. F., Maguire, D. J., and Rhind, D. W. (eds) (2005b) *Geographical Information Systems: Principles, Techniques, Management and Applications*. 2nd edn, abridged. Hoboken, NJ: John Wiley & Sons Ltd.

Malczewski, J. (1999) *GIS and Multicriteria Decision Analysis*. New York: John Wiley & Sons Ltd.

Malczewski, J. (2006) GIS-based multicriteria decision analysis: a survey of the literature. *International Journal of Geographical Information Science*, 20, 703–726.

Martin, D. (1989) Mapping population data from zone centroid locations. *Transactions of the Institute of British Geographers*, 14, 90–97.

Martin, D. (1996) *Geographic Information Systems: Socioeconomic Applications*. 2nd edn. London: Routledge.

Martin, D., Dorling, D. and Mitchell, R. (2002) Linking censuses through time: problems and solutions. *Area*, 34, 82–91.

Mather, P. M. (2004) *Computer Processing of Remotely-Sensed Images: An Introduction*. 3rd edn. Chichester: John Wiley & Sons Ltd.

Matheron, G. (1971) *The Theory of Regionalised Variables and its Applications*. Fontainebleau: Centre de Morphologie Mathématique de Fontainebleau.

McDonnell, R. and Kemp, K. (1995) *International GIS Dictionary*. Hoboken, NJ: John Wiley & Sons Ltd.

McGarigal, K. and Marks, B. J. (1995) *FRAGSTATS: Spatial Pattern Analysis Program for Quantifying Landscape Structure*. General Technical Report PNW-GTR-351. Portland, OR: US Department of Agriculture, Forest Service, Pacific Northwest Research Station.

Miller, H. J. (2008) Geographic data mining and knowledge discovery. In Wilson, J. P. and Fotheringham, A. S. (eds), *The Handbook of Geographic Information Science*. Maldon, MA: Blackwell Publishing, pp. 352–366.

Mitás, L. and Mitásová, H. (1999) Spatial interpolation. In Longley, P. A., Goodchild, M. F., Maguire, D. J., and Rhind, D. W. (eds), *Geographical Information Systems. Volume I: Principles and Technical Issues*. 2nd edn. New York: John Wiley & Sons Ltd, pp. 481–492.

Mitásová, H., Hofierka, J., Zlocha, M., and Iverson, L. R. (1996) Modelling topographic potential for erosion and deposition using GIS. *International Journal of Geographical Information Systems*, 10, 629–641.

Moran, P. A. P. (1950) Notes on continuous stochastic phenomena. *Biometrika*, 37, 17–23.

Mulla, D. J. (1988) Using geostatistics and spectral analysis to study spatial patterns in the topography of southeastern Washington state, U.S.A. *Earth Surface Processes and Landforms*, 13, 389–405.

Neteler, M. and Mitásová, H. (2007) *Open Source GIS: A GRASS GIS Approach*. 3rd edn. New York: Springer.

O'Brien, D. J., Kaneene, J. B., Getis, A., Lloyd, J. W., Swanson, G. M., and Leader, R. W. (2000) Spatial and temporal comparison of selected cancers in dogs and humans, Michigan, USA, 1964–1994. *Preventive Veterinary Medicine*, 47, 187–204.

Openshaw, S. (1984) *The Modifiable Areal Unit Problem*. Concepts and Techniques in Modern Geography 38. Norwich: GeoBooks.

Openshaw, S. and Taylor, P. J. (1979) A million or so correlation coefficients: three experiments on the modifiable areal unit problem. In Wrigley, N. (ed.), *Statistical Applications in the Spatial Sciences*. London: Pion, pp. 127–144.

Openshaw, S., Charlton, M. E., Wymer, C., and Craft, A. W. (1993) A mark I Geographical Analysis Machine for the automated analysis of point data sets. *International Journal of Geographical Information Systems*, 1, 359–377.

O'Sullivan, D. and Unwin, D. J. (2002) *Geographic Information Analysis*. Hoboken, NJ: John Wiley & Sons Ltd.

P

Pebesma E. J. (2004) Multivariable geostatistics in S: the Gstat package. *Computers and Geosciences*, 30, 683–691.

Pebesma E. J. and Wesseling, C. G. (1998) Gstat, a program for geostatistical modelling, prediction and simulation. *Computers and Geosciences*, 24, 17–31.

Peuker, T. K., Fowler, R. J., Little, J. J., and Mark. D. M. (1978) The triangulated irregular network. In American Society of Photogrammetry (ed.), *Proceedings of the Digital Terrain Models (DTM) Symposium. St. Louis, Missouri, May 9–11, 1978*. Falls Church, VA: American Society of Photogrammetry, pp. 516–540.

Pietro, L. S., O'Neal, M. A., and Puleo, J. A. (2008) Developing terrestrial-LiDAR-based digital elevation models for monitoring beach nourishment performance. *Journal of Coastal Research*, 24, 1555–1564.

Plane, D. A. and Rogerson, P. A. (1994) *The Geographical Analysis of Population: With Applications to Business and Planning*. Hoboken, NJ: John Wiley & Sons Ltd.

R

Rogerson, P. A. (2006) *Statistical Methods for Geography. A Student's Guide*. 2nd edn. London: SAGE Publications.

Rowntree, D. (2000) *Statistics without Tears: An Introduction for Non-Mathematicians*. London: Penguin.

S

Schabenberger, O. and Gotway, C. A. (2005) *Statistical Methods for Spatial Data Analysis*. Boca Raton, FL: Chapman and Hall/CRC Press.

Scott, D. M., Novak, D. C., Aultman-Hall, L., and Guo, F. (2006) Network Robustness Index: A new method for identifying critical links and evaluating the performance of transportation networks. *Journal of Transport Geography*, 14, 215–227.

Seeger, H. (2005) Spatial referencing and coordinate systems. In Longley, P. A., Goodchild, M. F., Maguire, D. J., and Rhind, D. W. (2005b) (eds), *Geographical Information Systems: Principles, Techniques, Management and Applications.* 2nd edn, abridged. Hoboken, NJ: John Wiley & Sons Ltd, pp. 427–436.

Shekhar, S., Zhang, P., Huang, Y., and Vatsavai, R. R. (2003) Trends in spatial data mining. In Kargupta, H., Joshi, A., Sivakumar, K., and Yesha, Y. (eds), *Data Mining: Next Generation Challenges and Future Directions.* Menlo Park, CA: AAAI/MIT Press, pp. 357–379.

Shennan, S. (1997) *Quantifying Archaeology.* 2nd edn. Edinburgh: Edinburgh University Press.

Silverman, B. W. (1986) *Density Estimation for Statistics and Data Analysis.* London: Chapman and Hall.

Sonka, M., Hlavac, V., and Boyle, R. (1999) *Image Processing, Analysis and Machine Vision.* 2nd edn. Pacific Grove, CA: PWS Publishing.

Sprague, D. S., Iwasaki, N., and Takahashi, S. (2007) Measuring rice paddy persistence spanning a century with Japan's oldest topographic maps: georeferencing the Rapid Survey Maps for GIS analysis. *International Journal of Geographical Information Science*, 21, 83–95

Steinberg, S. J. and Steinberg, S. L. (2006) *GIS. Geographical Information Systems for the Social Sciences: Investigating Space and Place.* Thousand Oaks, CA: Sage.

Taaffe, E. J., Gauthier, H. L., and O'Kelly, M. E. (1996) *Geography of Transportation.* 2nd edn. Englewood Cliffs, NJ: Prentice Hall.

Tate, N. J. and Atkinson, P. M. (eds) (2001) *Modelling Scale in Geographical Information Science.* Chichester: John Wiley & Sons Ltd.

Tate, N., Fisher, P. F., and Martin, D. J. (2008) Geographic information systems and surfaces. In Wilson, J. P. and Fotheringham, A. S. (eds), *The Handbook of Geographic Information Science.* Malden, MA: Blackwell, pp. 239–258.

Tobler, W. R. (1970) A computer movie simulating urban growth in the Detroit region. *Economic Geography*, 46, 234–240.

Tobler, W. R. (1979) Smooth pycnophylactic interpolation for geographical regions. *Journal of the American Statistical Association*, 74, 519–536.

Waller, L. A. and Gotway, C. A. (2004) *Applied Spatial Statistics for Public Health Data.* Hoboken, NJ: John Wiley & Sons Ltd.

Ward, M. D. and Gleditsch, K. S. (2008) *Spatial Regression Models.* Quantitative Applications in the Social Sciences 155. Los Angeles: Sage.

Webster, R. and Oliver, M. A. (2007) *Geostatistics for Environmental Scientists.* 2nd edn. Chichester: John Wiley & Sons Ltd.

Weisstein, E. W. (2003) *CRC Concise Encyclopedia of Mathematics.* 2nd edn. Boca Raton, FL: CRC Press.

Wheatley, D. W. (1995) Cumulative viewshed analysis: a GIS-based method for investigating intervisibility, and its archaeological application. In Lock, G. and Stancic, Z. (eds), *Archaeology and Geographical Information Systems: a European Perspective.* London: Taylor and Francis, pp. 171–186.

Wheeler, D. (2007) Diagnostic tools and a remedial method for collinearity in geographically weighted regression. *Environment and Planning A*, 39, 2464–2481.

Wheeler, D. and Tiefelsdorf, M. (2005) Multicollinearity and correlation among local regression coefficients in geographically weighted regression. *Journal of Geographical Systems*, 7, 161–187.

Wilson, A. G. and Kirkby, M. J. (1980) *Mathematics for Geographers and Planners.* 2nd edn. Oxford: Clarendon Press.

Wilson, J. P and Fotheringham, A. S. (eds) (2008) *The Handbook of Geographic Information Science.* Maldon, MA: Blackwell Publishing.

Wise, S. (2002) *GIS Basics.* London: Taylor and Francis.

Wong, D. W. S. (1997) Spatial dependency of segregation indices. *The Canadian Geographer*, 41, 128–136.

Worboys, M. and Duckman, M. (2004) *GIS: A Computing Perspective.* 2nd edn. Boca Raton, FL: CRC Press.

Appendix A
Matrix multiplication

The multiplication of a matrix by another matrix is illustrated in Figure A.1. In this case, the matrix has been multiplied by a copy of itself that has been flipped along its diagonal (this new matrix is called the transpose of the original matrix).

Each entry in the multiplied matrix is the sum of values in the cell's row in the original matrix multiplied by the values in the cell's column in the transpose of the matrix. For example, the value 251 (top-left cell) in the multiplied matrix is obtained from:

$$(1\times1)+(5\times5)+(9\times9)+(12\times12)=1+25+81+144=251$$

As another example, the value 309 (second row from the top, second column) of the multiplied matrix is obtained from:

$$(2\times2)+(6\times6)+(10\times10)+(13\times13)=4+36+100+169=309$$

The cells used in this example are highlighted in Figure A.2.

1	5	9	12
2	6	10	13
3	7	11	14
4	8	12	15

Original matrix

X

1	2	3	4
5	6	7	8
9	10	11	12
12	13	14	15

Flipped matrix
(transpose)

=

251	278	305	332
278	309	340	371
305	340	375	410
332	371	410	449

Multiplied matrix

Figure A.1 Matrix multiplication.

1	5	9	12
2	**6**	**10**	**13**
3	7	11	14
4	8	12	15

Original matrix

X

1	**2**	3	4
5	**6**	7	8
9	**10**	11	12
12	**13**	14	15

Flipped matrix
(transpose)

=

251	278	305	332
278	**309**	340	371
305	340	375	410
332	371	410	449

Multiplied matrix

Figure A.2 Matrix multiplication: selection of cells for the output cell in column 2, row 2.

Appendix B
The exponential function

One way to compute the exponential function for a particular value (following the example, -1.125) is to use the Maclaurin series for the exponential function (see Weisstein (2003) for more details):

$$\exp^x = 1 + -x + \frac{x^2}{2!} + \frac{x^3}{3!} + \frac{x^4}{4!} + \frac{x^5}{5!}\cdots$$

where the three dots indicate that other terms can be added with the numbers incremented (the next entry would read $x^6/6!$) and the more terms added in this way, the more accurate will be the result. The denominators in each term (e.g. 2!) are factorials. The factorial of 1 (i.e. 1!) is 1, the factorial of 2(2!) is 1×2, while $3!=1\times2\times3$, $4!=1\times2\times3\times4$, and so on. Using six terms this gives:

$$1 + -1.125 + \frac{-1.125^2}{2!} + \frac{-1.125^3}{3!} + \frac{-1.125^4}{4!} + \frac{-1.125^5}{5!} + \frac{-1.125^6}{6!}$$

$$= 1 + -1.125 + \frac{1.2656}{2} + \frac{-1.4238}{6} + \frac{1.6013}{24} + \frac{-1.8020}{120} + \frac{2.0273}{720}$$

$$= 1 + -1.125 + 0.6328 + -0.2373 + 0.0667 + -0.0150 + 0.0028$$

$$= 0.3250$$

Note the slight difference from the value computed in Section 8.4 (0.3247). This value was computed using the Microsoft® Excel® spreadsheet package 'exp' function and would be matched if more terms were added to the Maclaurin series (with eight terms the figures are the same to four decimal places).

Appendix C
The inverse tangent

You can obtain atan for a given value easily using standard spreadsheet functions, but if you want to compute atan yourself, the following equation can be used:

$$\text{atan}(g) = g - \frac{1}{3}g^3 + \frac{1}{5}g^5 - \frac{1}{7}g^7 + \cdots \quad \text{for } -1 < g < 1$$

This is called the Maclaurin series for the inverse tangent (Weisstein, 2003). It is assumed that g takes a value between -1 and 1 (indicated by 'for $-1<g<1$'). Notice the dots ($\cdots$), these indicate that we can add terms. This is easy to do once you recognize that there is a sequence: first there is a subtraction, then an addition, then a subtraction, then an addition, and so on. Also, the numbers step up by 2 from 3 (as in $1/3$ and g^3) to 5 and then 7. Put simply, the more terms added the more accurate the value of atan will be. The next term added would, following the sequence, entail addition and the value nine:

$$\text{atan}(g) = g - \frac{1}{3}g^3 + \frac{1}{5}g^5 - \frac{1}{7}g^7 + \frac{1}{9}g^9$$

Given the example in Section 10.5, the series was extended to a value of 11:

$$\text{atan}(0.5376) = 0.5376 - \frac{1}{3}0.5376^3 + \frac{1}{5}0.5376^5 - \frac{1}{7}0.5376^7 + \frac{1}{9}0.5376^9 - \frac{1}{11}0.5376^{11}$$

$$= 0.4933$$

Gradient in degrees is then given by:

$$gd = 0.4933 \times 57.29578 + 28.264°$$

Appendix D
Line intersection

Identification of line intersections is conceptually straightforward but computationally demanding. Line intersection is dealt with at length by Wise (2002), who outlines an algorithm for line intersection that computes outputs at various stages depending on whether the line or lines meet the specified criteria. The line intersection procedure is outlined below for two arcs. The first arc has the (start) coordinates x_1^s, y_1^s at one end and the (end) coordinates x_1^e, y_1^e at the other end. The second arc has the coordinates x_2^s, y_2^s at one end and the coordinates x_2^e, y_2^e at the other end. Note that the algorithm for line intersection is presented as what is called pseudocode. This means that the various stages are presented in a similar manner to computer code, in a series of logical steps.

If both lines are vertical (if so, the routine ends at this stage, as the lines can't intersect), i.e. if $x_1^e = x_1^s$ and $x_2^e = x_2^s$, then stop.

If the first line is vertical ($x_1^e = x_1^s$) then:

$$b_2 = \frac{y_2^e - y_2^s}{x_2^e - x_2^s}$$

$$a_2 = y_2^s - b_2 \times x_2^s$$

$$xp = x_1^s$$

$$yp = a_2 + b_2 \times xp$$

where xp and yp are the x and y coordinates where the lines intersect (following the calculations below and assuming various conditions are met).

Note that it is necessary that the algorithm tests for individual vertical lines as otherwise the result would be division by 0 (Wise, 2002) (e.g. if $x_1^e - x_1^s = 0$, b_1 cannot be calculated by $y_1^e - y_1^s / x_1^e - x_1^s$).

If the second line is vertical ($x_2^s = x_2^e$) then:

$$b_1 = \frac{y_1^e - y_1^s}{x_1^e - x_1^s}$$

$$a_1 = y_1^s - b_1 \times x_1^s$$

$$xp = x_2^s$$

$$yp = a_1 + b_1 \times xp$$

If neither line is vertical ($x_1^s \neq x_1^e$ and $x_2^s \neq x_2^e$) then:

$$b_1 = \frac{y_1^e - y_1^s}{x_1^e - x_1^s}$$

$$b_2 = \frac{y_2^e - y_2^s}{x_2^e - x_2^s}$$

$$a_1 = y_1^s - b_1 \times x_1^s$$

$$a_2 = y_2^s - b_2 \times x_2^s$$

If the lines are parallel (if so, the routine ends at this stage, as the lines can't intersect), that is if $b_1 = b_2$, then stop, else:

$$xp = -\frac{a_1 - a_2}{b_1 - b_2}$$

$$yp = a_1 + b_1 \times xp$$

Test if the intersection point falls on both lines, if so the intersection point is returned, that is:

if $(x_1^s - xp) \times (xp - x_1^e) \geq 0$ (this is test 1: t_1) and

$(x_2^s - xp) \times (xp - x_2^e) \geq 0$ (t_2) and

$(y_1^s - yp) \times (yp - y_1^e) \geq 0$ (t_3) and

$(y_2^s - yp) \times (yp - y_2^e) \geq 0$ (t_4)

then the lines cross at *xp, yp*. If all of these four conditions are not met then the lines do not intersect.

Note that ≥ means 'greater than or equal to'.

This procedure is illustrated using the three arcs shown in Figure D.1. Arc 2 is depicted using a different line symbol, so it is clearly differentiated from arc 1.

Obviously, arcs 1 and 2 do intersect while arc 3 does not intersect with either of the other two arcs. In such a simple case, it is easy to identify overlaps but if we have many arcs and we want to know exactly where they overlap (if they do) then we need some automated procedure, like the one outlined above.

Table D.1 gives the coordinates of the nodes for the arcs shown in Figure D.1, while Table D.2 gives the parameter values following the algorithm outlined above.

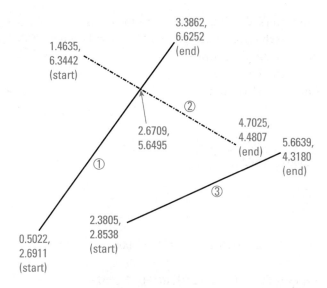

Figure D.1 Simple line features with *x, y* coordinates indicated.

Table D.1 Coordinates of nodes

Node	x	y
x_1^s, y_1^s	0.5022	2.6911
x_1^e, y_1^e	3.3862	6.6252
x_2^s, y_2^s	1.4635	6.3442
x_2^e, y_2^e	4.7025	4.4807
x_3^s, y_3^s	2.3805	2.8538
x_3^e, y_3^e	5.6639	4.3180

Table D.2 Parameter values for intersection calculations

| Parameter | Intersection test for: | |
	Arcs 1 and 2	Arcs 1 and 3
b_1	1.3641	1.3641
b_2	−0.5753	0.4459
a_1	2.0060	2.0060
a_2	7.1862	1.7922
xp	2.6709	−0.2329
yp	5.6495	1.6884
t_1	1.5512	−2.6602
t_2	2.4530	−15.4103
t_3	2.8865	−4.9501
t_4	0.8120	−3.0645

The algorithm is worked through here to test the intersection of arcs 1 and 2 (the values computed at each stage are also given in the second column of Table D.2):

Both lines not are vertical, **continue.**

First line not vertical, **continue.**

Second line not vertical, **continue.**

Neither line is vertical:

$$b_1 = \frac{y_1^e - y_1^s}{x_1^e - x_1^s} = \frac{6.6252 - 2.6911}{3.3862 - 0.5022} = 1.3641$$

$$b_2 = \frac{y_2^e - y_2^s}{x_2^e - x_2^s} = \frac{4.4807 - 6.3442}{4.7025 - 1.4635} = -0.5753$$

$$a_1 = y_1^s - (b_1 \times x_1^s) = 2.6911 - (1.3641 \times 0.5022) = 2.0060$$

$$a_2 = y_2^s - (b_2 \times x_2^s) = 6.3442 - (-0.5753 \times 1.4635) = 7.1862$$

Lines are not parallel, **continue**

$$xp = -\frac{a_1 - a_2}{b_1 - b_2} = -\frac{2.0060 - 7.1862}{1.3641 - -0.5753} = 2.6709$$

$$yp = a_1 + (b_1 \times xp) = 2.0060 + (1.3641 \times 2.6709) = 5.6495$$

Test if the intersection point falls on both lines.

$$t_1 = (x_1^s - xp) \times (xp - x_1^e) = (0.5022 - 2.6709) \times (2.6709 - 3.3862) = 1.5512$$

$$t_2 = (x_2^s - xp) \times (xp - x_2^e) = (1.4635 - 2.6709) \times (4.7025 - 3.3862) = 2.4530$$

$$t_3 = (y_1^s - yp) \times (yp - y_1^e) = (2.6911 - 5.6495) \times (5.6495 - 6.6252) = 2.8865$$

$$t_4 = (y_2^s - yp) \times (yp - y_2^e) = (6.3442 - 5.6495) \times (5.6495 - 4.4808) = 0.8120$$

All of the above four values are greater than 0, so the lines intersect at the location with $x = 2.6709$ and $y = 5.6495$ (as marked on Figure D.1).

To speed up the process of identifying intersections, a polygon minimum enclosing rectangle (MER) can be defined which bounds an arc. The MER is a rectangle that is of the minimum size necessary to contain the whole of an arc. If the MERs for two arcs do not intersect (i.e. overlap), then the arcs that they contain cannot intersect and there is no need to compare them. Comparison of MERs is straightforward and, by removing the need to compare complex arcs that cannot possibly overlap, the use of MERs may save much time. In the example here, the arcs are simple straight segments

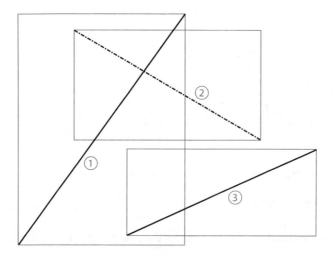

Figure D.2 MERs around simple line features.

whereas in most real-world applications this is unlikely to be the case. Figure D.2 shows MERs around each of the arcs from Figure D.1. Figure D.2 shows that arcs 2 and 3 cannot intersect as their MERs do not ovelap, so in this case testing for intersection is not required. Arcs 1 and 2, and 1 and 3, do have overlapping MERs, and so testing for intersection would be conducted in these cases. Recall that, while it is obvious arcs 1 and 3 do not intersect, if there are many arcs visual assessments of this kind will be impossible and an automated approach, like that outlined above, is essential. The same MER process can be used to test each pair of line segments (Wise, 2002). An additional stage is to split lines into monotonic sections, for example a section where as x increases y values also increase. Two such sections can only intersect once and after an intersection between segments of two monotonic sections is identified there is no need to search for others. In other words, only if a line segment turns back on itself can two line segments intersect more than once.

With these principles in place we have a workable method for line intersection and this provides the basis for the overlay operators that are detailed in Chapter 5.

Appendix E
Ordinary least squares

The solution for regression coefficients is given by:

$$\beta = (\mathbf{Y}^T\mathbf{Y})^{-1}\mathbf{Y}^T\mathbf{z}$$ (E.1)

Taking the component in brackets first, for the example given in Section 3.3, we have:

$$\mathbf{Y}^T\mathbf{Y} = \begin{bmatrix} 1 & 1 & 1 & 1 & 1 & 1 & 1 & 1 & 1 \\ 12 & 34 & 32 & 12 & 11 & 14 & 56 & 75 & 43 \end{bmatrix} \times \begin{bmatrix} 1 & 12 \\ 1 & 34 \\ 1 & 32 \\ 1 & 12 \\ 1 & 11 \\ 1 & 14 \\ 1 & 56 \\ 1 & 75 \\ 1 & 43 \end{bmatrix}$$

If you do not understand matrix multiplication *after* working through this box then go to Appendix A. The output of this operation will be a 2×2 matrix since there are 2×2 multiplications: the entries in the top row of $\mathbf{Y}^T$ multiplied by the entries in the left column of $\mathbf{Y}$, the entries in the top row of $\mathbf{Y}^T$ multiplied by the entries in the right column of $\mathbf{Y}$, the entries in the bottom row of $\mathbf{Y}^T$ multiplied by the entries in the left column of $\mathbf{Y}$, and the entries in the bottom row of $\mathbf{Y}^T$ multiplied by the entries in the right column of $\mathbf{Y}$. We will call the output matrix $\mathbf{C}$ and its entries can be labelled as follows:

$$\mathbf{C} = \begin{bmatrix} c_{11} & c_{12} \\ c_{21} & c_{22} \end{bmatrix}$$

The multiplication of the two matrices ($\mathbf{Y}$ and $\mathbf{Y}^T$) is conducted as follows:

$c_{11} =$ top row of $\mathbf{Y}^T$ multiplied by left column of $\mathbf{Y}$

$c_{12} =$ top row of $\mathbf{Y}^T$ multiplied by right column of $\mathbf{Y}$

$c_{21} =$ bottom row of $\mathbf{Y}^T$ multiplied by left column of $\mathbf{Y}$

$c_{22} =$ bottom row of $\mathbf{Y}^T$ multiplied by right column of $\mathbf{Y}$

For this example, this leads to:

$$c_{11} = (1 \times 1) + (1 \times 1) + (1 \times 1) + (1 \times 1) + (1 \times 1) + (1 \times 1) + (1 \times 1) + (1 \times 1) + (1 \times 1)$$

$$c_{12} = (1 \times 12) + (1 \times 34) + (1 \times 32) + (1 \times 12) + (1 \times 11) + (1 \times 14) + (1 \times 56) + (1 \times 75) + (1 \times 43)$$

$$c_{21} = (12 \times 1) + (34 \times 1) + (32 \times 1) + (12 \times 1) + (11 \times 1) + (14 \times 1) + (56 \times 1) + (75 \times 1) + (43 \times 1)$$

$$c_{22} = (12 \times 12) + (34 \times 34) + (32 \times 32) + (12 \times 12) + (11 \times 11) + (14 \times 14)$$
$$+ (56 \times 56) + (75 \times 75) + (43 \times 43)$$

This leads to:

$$C = \begin{bmatrix} 9 & 289 \\ 289 & 13395 \end{bmatrix}$$

The superscript -1 in Equation E.1 indicates that we take the inverse of the multiplied matrices. The inverse of the matrix C (C^{-1}) multiplied by C equals the identity matrix, I: CC^{-1}.

The identity matrix is simply a square matrix of 0s which has 1s along its diagonal. The identity matrix for the case of 5×5 entries is thus:

$$
\begin{matrix}
1 & 0 & 0 & 0 & 0 \\
0 & 1 & 0 & 0 & 0 \\
0 & 0 & 1 & 0 & 0 \\
0 & 0 & 0 & 1 & 0 \\
0 & 0 & 0 & 0 & 1
\end{matrix}
$$

For a 2×2 matrix, finding the inverse is straightforward. In this case, the inverse of the matrix (see O'Sullivan and Unwin (2002) for a further description) is obtained as follows:

For a matrix C, $\begin{bmatrix} c_1 & c_2 \\ c_3 & c_4 \end{bmatrix}$:

$$C^{-1} = \frac{1}{c_1 c_4 - c_2 c_3} \times \begin{bmatrix} c_4 & -c_2 \\ -c_3 & c_1 \end{bmatrix}$$

Using the example data, the inverse of $Y^T Y$ is obtained as follows:

$$(Y^T Y)^{-1} = \frac{1}{(9 \times 13395) - (289 \times 289)} \times \begin{bmatrix} 13395 & -289 \\ -289 & 9 \end{bmatrix} = 0.000027 \times \begin{bmatrix} 13395 & -289 \\ -289 & 9 \end{bmatrix}$$

The final component means that each value in the 2×2 matrix is multiplied by 0.000027 leading to:

$$= \begin{bmatrix} 0.36169 & -0.00780 \\ -0.00780 & 0.00024 \end{bmatrix}$$

Note that, following the definition of the inverse above:

$$\mathbf{I} = \begin{bmatrix} 9 & 289 \\ 289 & 13395 \end{bmatrix} \begin{bmatrix} 0.36169 & -0.00780 \\ -0.00780 & 0.00024 \end{bmatrix} = \begin{bmatrix} 1 & 0 \\ 0 & 1 \end{bmatrix}$$

The remaining part of the operation is given by:

$$\mathbf{Y}^T \mathbf{z} = \begin{bmatrix} 1 & 1 & 1 & 1 & 1 & 1 & 1 & 1 & 1 \\ 12 & 34 & 32 & 12 & 11 & 14 & 56 & 75 & 43 \end{bmatrix} \times \begin{bmatrix} 6 \\ 52 \\ 41 \\ 25 \\ 22 \\ 9 \\ 43 \\ 67 \\ 32 \end{bmatrix} = \begin{bmatrix} 297 \\ 12629 \end{bmatrix}$$

and the calculations using this operation are:

$$(1 \times 6) + (1 \times 52) + (1 \times 41) + (1 \times 25) + (1 \times 22) + (1 \times 9) + (1 \times 43) + (1 \times 67) + (1 \times 32) = 297$$

$$(12 \times 6) + (34 \times 52) + (32 \times 41) + (12 \times 25) + (11 \times 22) + (14 \times 9) + (56 \times 43) \\ + (75 \times 67) + (43 \times 32) = 12629$$

As shown in Section 3.3, the intercept and slope are then obtained from:

$$\beta = (\mathbf{Y}^T \mathbf{Y})^{-1} \mathbf{Y}^T \mathbf{z} = \begin{bmatrix} 0.36169 & -0.00780 \\ -0.00780 & 0.00024 \end{bmatrix} \times \begin{bmatrix} 297 \\ 12629 \end{bmatrix} = \begin{bmatrix} 8.871 \\ 0.751 \end{bmatrix}$$

Appendix F
Ordinary kriging system

The book explains the workings of the large majority of techniques it introduces. In that spirit, this appendix shows in full how to make a prediction using the widely used approach of ordinary kriging.

Section 9.7 outlined ordinary kriging, but solving of the OK system to obtain the weights and the Lagrange multiplier was not demonstrated. This section provides a fully worked example of solving the OK system, using the same data as in Section 9.7.

In matrix notation, the OK system can be written as (Equation 9.16):

$$\mathbf{K}\lambda = \mathbf{k}$$

Recall that $\mathbf{K}$ is the $n+1 \times n+1$ (with n nearest neighbours used for prediction) matrix of semivariances between each of the observations:

$$\mathbf{K} = \begin{bmatrix} \gamma(\mathbf{x}_1 - \mathbf{x}_1) & \cdots & \gamma(\mathbf{x}_1 - \mathbf{x}_n) & 1 \\ \vdots & \vdots & \vdots & \vdots \\ \gamma(\mathbf{x}_n - \mathbf{x}_1) & \cdots & \gamma(\mathbf{x}_n - \mathbf{x}_n) & 1 \\ 1 & \cdots & 1 & 0 \end{bmatrix}$$

λ are the OK weights and $\mathbf{k}$ are semivariances for the observations to the prediction location:

$$\lambda = \begin{bmatrix} \lambda_1 \\ \vdots \\ \lambda_n \\ \psi \end{bmatrix} \qquad \mathbf{k} = \begin{bmatrix} \gamma(\mathbf{x}_1 - \mathbf{x}_0) \\ \vdots \\ \gamma(\mathbf{x}_n - \mathbf{x}_0) \\ 1 \end{bmatrix}$$

The OK weights are obtained by multiplying the inverse of the data semivariance matrix by the vector of data to prediction semivariances:

$$\lambda = \mathbf{K}^{-1}\mathbf{k}$$

The OK variance is then obtained from:

$$\sigma_{OK}^2 = \mathbf{k}^T\lambda$$

To obtain the weights we therefore need to obtain the inverse of the **K** matrix. Appendix E shows how to obtain the inverse of a 2×2 matrix. The matrix we are dealing with here has 5×5 entries, and the same approach will not work. This section shows how such a matrix may be inverted and thus the OK weights obtained.

For the example in Section 9.7.2 the OK system is given as:

$$\begin{bmatrix} 0 & 376.905 & 359.589 & 379.853 & 1 \\ 376.905 & 0 & 307.108 & 394.601 & 1 \\ 359.589 & 307.108 & 0 & 448.401 & 1 \\ 379.853 & 394.601 & 448.401 & 0 & 1 \\ 1 & 1 & 1 & 1 & 0 \end{bmatrix} \times \begin{bmatrix} \lambda_1 \\ \lambda_2 \\ \lambda_3 \\ \lambda_4 \\ \psi \end{bmatrix} = \begin{bmatrix} 268.116 \\ 311.250 \\ 311.983 \\ 367.662 \\ 1 \end{bmatrix}$$

To obtain the inverse of a matrix by hand, several prior stages must be worked through. Fortunately, computer software packages exist to do this very quickly.

First, we need to obtain the determinant of the matrix **K** (see Weisstein (2003) for a definition). Manually, this is a tedious process, but the principles will be demonstrated. The determinant of the matrix will be found using a method called expansion by minors. This method is appropriate for small matrices but other approaches such as Gaussian elimination are more efficient for large matrices (Weisstein, 2003).

The method of expansion by minors is based on splitting the matrix into smaller submatrices and obtaining the determinant of each submatrix. In this case, the original 5×5 matrix is split into five 4×4 matrices. Each of these five 4×4 matrices is then split into four 3×3 matrices. Finally, each of the 3×3 matrices is split into three 2×2 matrices and the determinants (denoted by Det) of each of these calculated as follows.

For a matrix **C**, $\begin{bmatrix} c_{11} & c_{12} \\ c_{21} & c_{22} \end{bmatrix}$:

$$\text{Det } \mathbf{C} = |\mathbf{C}| = (c_{11} \times c_{22}) - (c_{21} \times c_{12})$$

Using the method of expansion by minors, the determinant of a 3×3 matrix is given by:

$$\text{Det } \mathbf{C} = |\mathbf{C}| = \begin{vmatrix} c_{11} & c_{12} & c_{13} \\ c_{21} & c_{22} & c_{23} \\ c_{31} & c_{32} & c_{33} \end{vmatrix} = c_{11}\begin{vmatrix} c_{22} & c_{23} \\ c_{32} & c_{33} \end{vmatrix} - c_{12}\begin{vmatrix} c_{21} & c_{23} \\ c_{31} & c_{33} \end{vmatrix} + c_{13}\begin{vmatrix} c_{21} & c_{22} \\ c_{31} & c_{32} \end{vmatrix}$$

If the 3×3 matrix was in turn part of a larger matrix then its determinant would be derived as follows. This process is illustrated using the matrix **K**, which has 5×5 elements:

$$\begin{bmatrix} 0 & 376.905 & 359.589 & 379.853 & 1 \\ 376.905 & 0 & 307.108 & 394.601 & 1 \\ 359.589 & 307.108 & 0 & 448.401 & 1 \\ 379.853 & 394.601 & 448.401 & 0 & 1 \\ 1 & 1 & 1 & 1 & 0 \end{bmatrix}$$

Firstly, the top-left element is taken and its row and column are masked (the entries in that row and column are removed). This leads to:

$$\begin{bmatrix} 0 & 307.108 & 394.601 & 1 \\ 307.108 & 0 & 448.401 & 1 \\ 394.601 & 448.401 & 0 & 1 \\ 1 & 1 & 1 & 0 \end{bmatrix}$$

Note that, as in the example above, we multiply the determinant by the value that defined the masking row and column. Here the determinant is multiplied by 0, resulting in 0. We could therefore avoid this step. But, it is worked through for illustrative purposes.

We repeat the masking (removal) process for the new 4×4 matrix. Taking the top-left element and masking its row and column:

$$\begin{bmatrix} 0 & 448.401 & 1 \\ 448.401 & 0 & 1 \\ 1 & 1 & 0 \end{bmatrix}$$

Taking the top-left element and masking its row and column, we follow the steps outlined above for obtaining the determinant:

$$0 \times \begin{bmatrix} 0 & 1 \\ 1 & 0 \end{bmatrix} = 0((0 \times 0) - (1 \times 1)) = 0$$

This appears in Table F.1 as the first entry ($4 \times 4 = 1$, $3 \times 3 = 1$, $2 \times 2 = 1$, 3×3 element $= 0$) with:

$$\begin{bmatrix} c_{11} & c_{12} \\ c_{21} & c_{22} \end{bmatrix} = \begin{bmatrix} 0 & 1 \\ 1 & 0 \end{bmatrix}$$

Following this stage, we take the top middle element and mask its row and column:

$$448.401 \times \begin{bmatrix} 448.401 & 1 \\ 1 & 0 \end{bmatrix} = 448.401 \times ((448.401 \times 0) - (1 \times 1)) = -448.401$$

Table F.1 Determinants for the 2×2 matrices

4×4	3×3	2×2	3×3 element	c_{11}	c_{22}	c_{21}	c_{12}	3×3 element× $(c_{11} \times c_{22}) - (c_{21} \times c_{12})$
1	1	1	0	0	0	1	1	0.000
1	1	2	448.401	448.401	0	1	1	−448.401
1	1	3	1	448.401	1	1	0	448.401
1	2	4	307.108	0	0	1	1	−307.108
1	2	5	448.401	394.601	0	1	1	−448.401
1	2	6	1	394.601	1	1	0	394.601
1	3	7	307.108	448.401	0	1	1	−307.108
1	3	8	0	394.601	0	1	1	0.000
1	3	9	1	394.601	1	1	448.401	−53.800
1	4	10	307.108	448.401	1	1	0	137707.534
1	4	11	0	394.601	1	1	0	0.000
1	4	12	448.401	394.601	1	1	448.401	−24123.974
2	5	13	0	0	0	1	1	0.000
2	5	14	448.401	448.401	0	1	1	−448.401
2	5	15	1	448.401	1	1	0	448.401
2	6	16	359.589	0	0	1	1	−359.589
2	6	17	448.401	379.853	0	1	1	−448.401
2	6	18	1	379.853	1	1	0	379.853
2	7	19	359.589	448.401	0	1	1	−359.589
2	7	20	0	379.853	0	1	1	0.000
2	7	21	1	379.853	1	1	448.401	−68.548
2	8	22	359.589	448.401	1	1	0	161240.067
2	8	23	0	379.853	1	1	0	0.000
2	8	24	448.401	379.853	1	1	448.401	−30736.992
3	9	25	307.108	0	0	1	1	−307.108
3	9	26	448.401	394.601	0	1	1	−448.401
3	9	27	1	394.601	1	1	0	394.601
3	10	28	359.589	0	0	1	1	−359.589
3	10	29	448.401	379.853	0	1	1	−448.401
3	10	30	1	379.853	1	1	0	379.853
3	11	31	359.589	394.601	0	1	1	−359.589
3	11	32	307.108	379.853	0	1	1	−307.108
3	11	33	1	379.853	1	1	394.601	−14.748
3	12	34	359.589	394.601	1	1	0	141894.179
3	12	35	307.108	379.853	1	1	0	116655.895
3	12	36	448.401	379.853	1	1	394.601	−6613.018
4	13	37	307.108	448.401	0	1	1	−307.108
4	13	38	0	394.601	0	1	1	0.000
4	13	39	1	394.601	1	1	448.401	−53.800
4	14	40	359.589	448.401	0	1	1	−359.589
4	14	41	0	379.853	0	1	1	0.000

continued

Table F.1 *continued*

4×4	3×3	2×2	3×3 element	c_{11}	c_{22}	c_{21}	c_{12}	3×3 element × $(c_{11} \times c_{22}) - (c_{21} \times c_{12})$
4	14	42	1	379.853	1	1	448.401	−68.548
4	15	43	359.589	394.601	0	1	1	−359.589
4	15	44	307.108	379.853	0	1	1	−307.108
4	15	45	1	379.853	1	1	394.601	−14.748
4	16	46	359.589	394.601	1	1	448.401	−19345.888
4	16	47	307.108	379.853	1	1	448.401	−21051.639
4	16	48	0	379.853	1	1	394.601	0.000
5	17	49	307.108	448.401	1	1	0	137707.534
5	17	50	0	394.601	1	1	0	0.000
5	17	51	448.401	394.601	1	1	448.401	−24123.974
5	18	52	359.589	448.401	1	1	0	161240.067
5	18	53	0	379.853	1	1	0	0.000
5	18	54	448.601	379.853	1	1	448.401	−30750.701
5	19	55	359.589	394.601	1	1	0	141894.179
5	19	56	307.108	379.853	1	1	0	116655.895
5	19	57	448.401	379.853	1	1	394.601	−6613.018
5	20	58	359.589	394.601	1	1	448.401	−19345.888
5	20	59	307.108	379.853	1	1	448.401	−21051.639
5	20	60	0	379.853	1	1	394.601	0.000

This corresponds to the second entry in Table F.1. Next, we take the top-right element and mask its row and column:

$$1 \times \begin{bmatrix} 448.401 & 0 \\ 1 & 1 \end{bmatrix} = 1 \times ((448.401 \times 1) - (1 \times 0)) = 448.401$$

This corresponds to the third entry in Table F.1. The same procedure is worked through for the other matrices in turn. The second 3×3 submatrix and then the 2×2 submatrices derived from this follow:

$$\begin{bmatrix} 307.108 & 0 & 448.401 \\ 394.601 & 448.401 & 0 \\ 1 & 1 & 1 \end{bmatrix}$$

$$307.108 \times \begin{bmatrix} 0 & 1 \\ 1 & 0 \end{bmatrix} = 307.108 \times ((0 \times 0) - (1 \times 1)) = -307.108$$

$$448.401 \times \begin{bmatrix} 394.601 & 1 \\ 1 & 0 \end{bmatrix} = 448.401 \times ((394.601 \times 0) - (1 \times 1)) = -448.401$$

$$1 \times \begin{bmatrix} 394.601 & 0 \\ 1 & 1 \end{bmatrix} = 1 \times ((394.601 \times 1) - (1 \times 0)) = 394.601$$

The third 3×3 matrix…

$$\begin{bmatrix} 307.108 & 0 & 1 \\ 394.601 & 448.401 & 1 \\ 1 & 1 & 0 \end{bmatrix}$$

$$307.108 \times \begin{bmatrix} 448.401 & 1 \\ 1 & 0 \end{bmatrix} = 307.108 \times ((448.401 \times 0) - (1 \times 1)) = -307.108$$

$$0 \times \begin{bmatrix} 394.601 & 1 \\ 1 & 0 \end{bmatrix} = 0 \times ((394.601 \times 0) - (1 \times 1)) = 0$$

$$1 \times \begin{bmatrix} 394.601 & 448.401 \\ 1 & 1 \end{bmatrix} = 1 \times ((394.601 \times 1) - (1 \times 448.401)) = -53.800$$

The fourth 3×3 matrix…

$$\begin{bmatrix} 307.108 & 0 & 448.401 \\ 394.601 & 448.401 & 0 \\ 1 & 1 & 1 \end{bmatrix}$$

$$307.108 \times \begin{bmatrix} 448.401 & 0 \\ 1 & 1 \end{bmatrix} = 307.108 \times ((448.401 \times 1) - (1 \times 0)) = 137707.534$$

$$0 \times \begin{bmatrix} 394.601 & 0 \\ 1 & 1 \end{bmatrix} = 0 \times ((394.601 \times 1) - (1 \times 0)) = 0$$

$$448.401 \times \begin{bmatrix} 394.601 & 448.401 \\ 1 & 1 \end{bmatrix} = 448.401 \times ((394.601 \times 1) - (1 \times 448.401)) = -24123.974$$

The workings for the other four 4×4 matrices and the 3×3 matrices derived from these, with 2×2 matrices derived from the latter, are not given in full. The calculations are summarized in Table F.1 (2×2 matrices) and Table F.2 (3×3 and 4×4 matrices).

Table F.2 Determinants for the 3×3 and 4×4 matrices: D1, D2, and D3 are 3×3 elements multiplied by determinants of 2×2 matrices (see Table F.1)

4×4	3×3	D1	D2	D3	Sign 1	4×4 element	D1−D2+D3	Sign 1×4×4 element×(D1−D2+D3)=Col 9	Sign 2	5×5 element	Sum Col 9	Sign 2×5×5 element×Sum Col 9
1	1	0.000	−448.401	448.401	1	0	896.802	0.000				
1	2	−307.108	−448.401	394.601	−1	307.108	535.894	−164577.335				
1	3	−307.108	0.000	−53.800	1	394.601	−360.908	−142414.658				
1	4	137707.534	0.000	−24123.974	−1	1	113583.561	−113583.561	1	0.000	−420575.553	0.000
2	5	0.000	−448.401	448.401	1	376.905	896.802	338009.158				
2	6	−359.589	−448.401	379.853	−1	307.108	468.665	−143930.771				
2	7	−359.589	0.000	−68.548	1	394.601	−428.137	−168943.288				
2	8	161240.067	0.000	−30736.992	−1	1	130503.075	−130503.075	−1	376.905	−105367.977	39713717.291
3	9	−307.108	−448.401	394.601	1	376.905	535.894	201981.128				
3	10	−359.589	−448.401	379.853	−1	0	468.665	0.000				
3	11	−359.589	−307.108	−14.748	1	394.601	−67.229	−26528.631				
3	12	141894.179	116655.895	−6613.018	−1	1	18625.266	−18625.266	1	359.589	156827.232	56393347.356
4	13	−307.108	0.000	−53.800	1	376.905	−360.908	−136028.030				
4	14	−359.589	0.000	−68.548	−1	0	−428.137	0.000				
4	15	−359.589	−307.108	−14.748	1	307.108	−67.229	−20646.564				
4	16	−19345.888	−21051.639	0.000	−1	1	1705.751	−1705.751	−1	379.853	−158380.344	60161248.983
5	17	137707.534	0.000	−24123.974	1	376.905	113583.561	42810211.873				
5	18	161240.067	0.000	−30750.701	−1	0	130489.366	0.000				
5	19	141894.179	116655.895	−6613.018	1	307.108	18625.266	5719968.165				
5	20	−19345.888	−21051.639	0.000	−1	394.601	1705.751	−673091.044	1	1.000	47857088.994	47857088.994
											47857088.994	204125402.625
											Sum (Det **K**)	

As indicated above, Table F.1 gives numbers for each 4×4, 3×3, and 2×2 matrix (in the same order as presented above for the first 4×4 matrix). The element of the 3×3 matrix (i.e. the value of the 3×3 matrix that defines the masking row and column in a given case) is then given. The four elements in each 2×2 matrix are given by c_{11}, c_{22}, c_{21}, and c_{22}. The final column is the 3×3 element $\times (c_{11} \times c_{22}) - (c_{21} \times c_{12})$. In words, this is the element of the 3×3 matrix that defines the masking row and column multiplied by the determinant of the 2×2 matrix. These products appear in Table F.2 in columns D1, D2, and D3. Table F.2 shows how the determinants of the submatrices are multiplied by the appropriate sign (+ or −) and by the element defining the masking row and column in the (sub)matrix from which the submatrix is derived—that is, once the 2×2 matrices are derived their determinants are calculated and each is multiplied by the element in the 3×3 matrix that defined the masking row and column. There are three sets of such values for each 3×3 matrix (which we will call D1, D2, and D3, from the final column in Table F.1). We multiply the correct sign (in the column headed 'Sign 1' in Table F.2) by the 4×4 element defining the masking row and column for each 3×3 matrix and then by D1 − D2 + D3. This leads to the ninth column in Table F.2. Next the appropriate sign ('Sign 2' in Table F.2) is multiplied by the 5×5 element defining the masking row and column of each 4×4 matrix and then by the sum of products in the ninth column of Table F.2 for each 4×4 matrix. This gives the final column in Table F.2. The sum of values in the final column of Table F.2 is 204125402.625, and this is the determinant of **K**. It will be clear why we use computers to do this kind of thing!

As the next stage, we compute what is called the matrix of minors. This is the determinant for each cell obtained by masking the row and column of that cell. For example, for **K**:

$$
\begin{bmatrix}
0 & 376.905 & 359.589 & 379.853 & 1 \\
376.905 & 0 & 307.108 & 394.601 & 1 \\
359.589 & 307.108 & 0 & 448.401 & 1 \\
379.853 & 394.601 & 448.401 & 0 & 1 \\
1 & 1 & 1 & 1 & 0
\end{bmatrix}
$$

Taking the top-left element and masking its row and column (as before):

$$
\begin{bmatrix}
0 & 307.108 & 394.601 & 1 \\
307.108 & 0 & 448.401 & 1 \\
394.601 & 448.401 & 0 & 1 \\
1 & 1 & 1 & 0
\end{bmatrix}
$$

The determinant of this 4×4 matrix, obtained as before, is −420575.553. This is written to the top-left cell of the new matrix of minors and the same process is completed for all other cells. One more example is given for clarity. For the second cell down and across (with a value of zero) the rows and columns are masked as follows:

$$
\begin{bmatrix}
0 & \mathbf{376.905} & 359.589 & 379.853 & 1 \\
\mathbf{376.905} & 0 & \mathbf{307.108} & \mathbf{394.601} & \mathbf{1} \\
359.589 & \mathbf{307.108} & 0 & 448.401 & 1 \\
379.853 & \mathbf{394.601} & 448.401 & 0 & 1 \\
1 & \mathbf{1} & 1 & 1 & 0
\end{bmatrix}
$$

Removing this row and column leads to:

$$
\begin{bmatrix}
0 & 359.589 & 379.853 & 1 \\
359.589 & 0 & 448.401 & 1 \\
379.853 & 448.401 & 0 & 1 \\
1 & 1 & 1 & 0
\end{bmatrix}
$$

The determinant of this 4×4 matrix is calculated exactly as before using the method of expansion by minors. The value obtained, -461658.978, is added to the matrix of minors and the process is conducted for all rows and columns. The matrix of minors is:

$$
\begin{bmatrix}
-420575.553 & -105367.977 & 156827.232 & -158380.344 & 47857088.994 \\
-105367.977 & -461658.978 & -220804.585 & 135486.416 & -43137755.532 \\
156827.232 & -220804.585 & -441516.287 & -63884.471 & 51275340.249 \\
-158380.344 & 135486.416 & -63884.471 & -357751.231 & -61855217.850 \\
47857088.994 & -43137755.532 & 51275340.249 & -61855217.850 & -58192774139.342
\end{bmatrix}
$$

Once this has been done for each cell the signs of each element in the matrix of minors must be set as follows:

$$
\begin{bmatrix}
+ & - & + & - & + \\
- & + & - & + & - \\
+ & - & + & - & + \\
- & + & - & + & - \\
+ & - & + & - & +
\end{bmatrix}
$$

It is this ordering that defines the signs used in the determinant calculations above (see Table F.2, Sign 1 and Sign 2 columns).

In words, the top-left element in the matrix of minors is multiplied by 1, the second element from left in the top row is multiplied by −1 and so on. This leads to what is called the matrix of cofactors:

$$
\begin{bmatrix}
-420575.553 & 105367.977 & 156827.232 & 158380.344 & 47857088.994 \\
105367.977 & -461658.978 & 220804.585 & 135486.416 & 43137755.532 \\
156827.232 & 220804.585 & -441516.287 & 63884.471 & 51275340.249 \\
158380.344 & 135486.416 & 63884.471 & -357751.231 & 61855217.850 \\
47857088.994 & 43137755.532 & 51275340.249 & 61855217.850 & -58192774139.342
\end{bmatrix}
$$

Next, we need the adjoint (transpose) of the matrix of cofactors. However, the matrix is symmetric (the entry of row 3, column 2 is the same as the entry in row 2, column 3), so no further step is necessary.

Now we have the determinant and the matrix of minors we can generate the inverse of **K**. This is obtained by dividing each element in the adjoint of the matrix of cofactors by the determinant (204125402.625). For example, the top-left element of the matrix of cofactors is −420575.553. Dividing this value by 204125402.625 gives −0.002060. Following this process for all elements of the matrix of cofactors the inverse of **K** (K^{-1}) is:

$$
\begin{bmatrix}
-0.002060 & 0.000516 & 0.000768 & 0.000776 & 0.234450 \\
0.000516 & -0.002262 & 0.001082 & 0.000664 & 0.211330 \\
0.000768 & 0.001082 & -0.002163 & 0.000313 & 0.251195 \\
0.000776 & 0.000664 & 0.000313 & -0.001753 & 0.303026 \\
0.234449 & 0.211330 & 0.251195 & 0.303026 & -285.083451
\end{bmatrix}
$$

We can now compute the weights with:

$$\lambda = K^{-1}k$$

In this case, this leads to:

$-0.002060 \times 268.116 + 0.000516 \times 311.250 + 0.000768 \times 311.983 + 0.000776$
$\times 367.662 + 0.234450 \times 1 = 0.36765$

$-0.000516 \times 268.116 - 0.002262 \times 311.250 + 0.001082 \times 311.983 + 0.000664$
$\times 367.662 + 0.211330 \times 1 = 0.22730$

$0.000768 \times 268.116 - 0.001082 \times 311.250 - 0.002163 \times 311.983 + 0.000313$
$\times 367.662 + 0.251195 \times 1 = 0.23413$

$0.000776 \times 268.116 - 0.000664 \times 311.250 - 0.000313 \times 311.983 - 0.001753$
$\times 367.662 + 0.303026 \times 1 = 0.17092$

$0.234449 \times 268.116 + 0.211330 \times 311.250 + 0.251195 \times 311.983 + 0.303026$
$\times 367.662 - 285.083451 \times 1 = 33.33221$

This gives us the four weights and the Lagrange multiplier. The weights matrix is therefore:

$$\lambda = \begin{bmatrix} 0.36765 \\ 0.22730 \\ 0.23413 \\ 0.17092 \\ 33.33221 \end{bmatrix}$$

where the top value is λ_1, the penultimate value is λ_4, and the last value is ψ, the Lagrange multiplier.

The predicted value is then given by (with values rounded to three decimal places):

$$(0.368 \times 68) + (0.227 \times 29) + (0.234 \times 48) + (0.171 \times 53) = 51.889$$

Using the Lagrange multiplier, the kriging variance is given by:

$$(0.368 \times 268.116) + (0.227 \times 311.250) + (0.234 \times 311.983) + (0.171 \times 367.662) \\ + (33.332 \times 1) = 338.537$$

The process is clearly non-trivial, but OK is implemented in many GIS such as ArcGIS™, where hundreds of predictions can be made in seconds.

Appendix G
Common problems in spatial data analysis

The following table provides a short list of some key spatial analysis tasks with details of appropriate methods and sections in which relevant material can be found.

Table G.1 Problems and corresponding methods

Problem	Methods	Chapter/s	Section/s
Alter zonal system	Areal interpolation	9	9.9
Characterize network complexity	Various indices	6	6.4
Characterize point pattern structure	Nearest-neighbour methods, K function	7	7.4
Combine multiple criteria	Multicriteria decision analysis	5	5.3
Compute cost surface	Cost surface (friction) analysis	10	10.6
Compute areas of indivisibility	Viewshed analysis	10	10.6
Compute moving window statistics	Various moving window procedures	4, 8, 10	4.6, 8.5.2, 10.4
Compute elevation derivatives (gradient, etc.)	Various procedures	10	10.5
Predict missing values	Spatial interpolation	9	9.1–9.10
Explore (spatial) relationships	Global regression, moving window and geographically weighted regression	8	8.5, 8.7.2
Extract a data subset	Spatial querying (e.g. using structured query language)	2	2.11
Identify the shortest route along a network	Shortest path algorithm	6	6.5
Make inferences about a population from a sample	Inferential statistics	3	3.4–3.5
Map point event intensity	Quadrats, kernel estimation	7	7.3
Measure lengths	Length algorithm	4	4.3
Measure areas	Area algorithm	4	4.4
Measure spatial autocorrelation	Various spatial autocorrelation measures	4, 8	4.8, 8.2, 8.4.1, 8.7.1
Overlay data layers	Various overlay procedures	5	5.1–5.4
Smooth image or enhance contrast	Various image processing procedures	10	10.3–10.4
Summarize numerical values	Summary statistics	3	3.1–3.3

Index